# Futuristic Design and Intelligent Computational Techniques in Neuroscience and Neuroengineering

Harjit Pal Singh
*CT Institute of Engineering and Technology, Punjab, India*

Anurag Sharma
*GNA University, Punjab, India*

Vikas Khullar
*CT Institute of Engineering and Technology, Punjab, India*

A volume in the Advances in Medical Technologies and Clinical Practice (AMTCP) Book Series

IGI Global
PUBLISHER of TIMELY KNOWLEDGE

Published in the United States of America by
IGI Global
Medical Information Science Reference (an imprint of IGI Global)
701 E. Chocolate Avenue
Hershey PA, USA 17033
Tel: 717-533-8845
Fax: 717-533-8661
E-mail: cust@igi-global.com
Web site: http://www.igi-global.com



Library of Congress Cataloging-in-Publication Data

Names: Singh, Harjit Pal, 1983- editor. | Sharma, Anurag, 1984- editor. | Khullar, Vikas, 1983- editor.
Title: Futuristic design and intelligent computational techniques in neuroscience and neuroengineering / Harjit Pal Singh, Anurag Sharma, and Vikas Khullar, editor.
Description: Hershey PA : Engineering Science Reference, [2021] | Includes bibliographical references and index. | Summary: "This research book include quality chapters on computational models, designs and multidisciplinary approaches for neurological diagnosis and treatment, offering a resource of neurological databases, computational intelligence, brain health informatics, effective analysis of neural functions and technological interventions"-- Provided by publisher.
Identifiers: LCCN 2021000710 (print) | LCCN 2021000711 (ebook) | ISBN 9781799874331 (hardcover) | ISBN 9781799874348 (ebook)
Subjects: LCSH: Nervous system--Diseases--Diagnosis--Data processing. | Nervous system--Diseases--Treatment--Data processing. | Biomedical engineering--Data processing. | Neuroinformatics. | Computational neuroscience.
Classification: LCC RC348 .H297 2021 (print) | LCC RC348 (ebook) | DDC 616.800285--dc23
LC record available at https://lccn.loc.gov/2021000710
LC ebook record available at https://lccn.loc.gov/2021000711

This book is published in the IGI Global book series Advances in Medical Technologies and Clinical Practice (AMTCP) (ISSN: 2327-9354; eISSN: 2327-9370)

British Cataloguing in Publication Data
A Cataloguing in Publication record for this book is available from the British Library.
All work contributed to this book is new, previously-unpublished material.
The views expressed in this book are those of the authors, but not necessarily of the publisher.

For electronic access to this publication, please contact: eresources@igi-global.com.

# Advances in Medical Technologies and Clinical Practice (AMTCP) Book Series

ISSN:2327-9354
EISSN:2327-9370



## MISSION

Medical technological innovation continues to provide avenues of research for faster and safer diagnosis and treatments for patients. Practitioners must stay up to date with these latest advancements to provide the best care for nursing and clinical practices.

The **Advances in Medical Technologies and Clinical Practice (AMTCP) Book Series** brings together the most recent research on the latest technology used in areas of nursing informatics, clinical technology, biomedicine, diagnostic technologies, and more. Researchers, students, and practitioners in this field will benefit from this fundamental coverage on the use of technology in clinical practices.

## COVERAGE

- Clinical Nutrition
- Clinical Studies
- Nursing Informatics
- Biometrics
- Clinical Data Mining
- Neural Engineering
- Biomechanics
- Patient-Centered Care
- E-Health
- Biomedical Applications



The Advances in Medical Technologies and Clinical Practice (AMTCP) Book Series (ISSN 2327-9354) is published by IGI Global, 701 E. Chocolate Avenue, Hershey, PA 17033-1240, USA, www.igi-global.com. This series is composed of titles available for purchase individually; each title is edited to be contextually exclusive from any other title within the series. For pricing and ordering information please visit http://www.igi-global.com/book-series/advances-medical-technologies-clinical-practice/73682. Postmaster: Send all address changes to above address. 

## Titles in this Series

*For a list of additional titles in this series, please visit:*
*http://www.igi-global.com/book-series/advances-medical-technologies-clinical-practice/73682*

***Assistive Technologies for Assessment and Recovery of Neurological Impairments***
Fabrizio Stasolla (University Giustino Fortunato of Benevento, Italy)
Medical Information Science Reference • © 2022 • 396pp • H/C (ISBN: 9781799874300) • US $295.00

***Ethical Implications of Reshaping Healthcare With Emerging Technologies***
Thomas Heinrich Musiolik (Berlin University of the Arts, Germany) and Alexiei Dingli (University of Malta, Malta)
Medical Information Science Reference • © 2022 • 242pp • H/C (ISBN: 9781799878889) • US $295.00

***Innovative Approaches for Nanobiotechnology in Healthcare Systems***
Touseef Amna (Albaha University, Saudi Arabia) and M. Shamshi Hassan (Albaha University, Saudi Arabia)
Medical Information Science Reference • © 2022 • 424pp • H/C (ISBN: 9781799882510) • US $395.00

***Machine Learning and Data Analytics for Predicting, Managing, and Monitoring Disease***
Manikant Roy (Lovely Professional University, India) and Lovi Raj Gupta (Lovely Professional University, India)
Medical Information Science Reference • © 2021 • 241pp • H/C (ISBN: 9781799871880) • US $395.00

***Advancing the Investigation and Treatment of Sleep Disorders Using AI***
M. Rajesh Kumar (Vellore Institute of Technology, Vellore, India) Ranjeet Kumar (Vellore Institute of Technology, Chennai, India) and D. Vaithiyanathan (National Institute of Technology, Delhi, India)
Medical Information Science Reference • © 2021 • 290pp • H/C (ISBN: 9781799880189) • US $345.00

*For an entire list of titles in this series, please visit:*
*http://www.igi-global.com/book-series/advances-medical-technologies-clinical-practice/73682*

IGI Global
PUBLISHER of TIMELY KNOWLEDGE

701 East Chocolate Avenue, Hershey, PA 17033, USA
Tel: 717-533-8845 x100 • Fax: 717-533-8661
E-Mail: cust@igi-global.com • www.igi-global.com

# Table of Contents

# Detailed Table of Contents



This chapter presents the P300-based human machine interference (HMI) systems control robotic wheel chair (RWC) prototype in right, left, forward, backward, and stop positions. Four different targets letters are used to elicit the P300 waves, flickering in the low frequency region, by using oddball paradigms and displayed on a liquid crystal display (LCD) screen by Lab-VIEW. After the pre-processing and taking one second time window, feature is extracted by using discrete wavelet transform (DWT). Three different classifiers—two based on ANNs pattern recognition neural network (PRNN) and feed forward neural network (FFNN) and the and other one based on support vector machine (SVM)—are used. Those three techniques are designed and compared with the different accuracies among them.



This study aimed to investigate the impact of game-based learning on student performance. For this reason, an empirical study was conducted which comprises the comparison of traditional learning and game-based learning. The participants were lower primary school students of age group 6-8 years. GSR NUL-217 logger sensor was used to record the physiological signals of each participant in real time. An Android-based game intervention named "KidsZoneApp" was developed which included mathematics and English lesson plans according to the student's curriculum. The collected log data was used to calculate the changes in different dimensions of completing the task. ANOVA yielded a very significant difference between game-based learning and traditional learning groups. Overall, the results showed that game-based learning contributed to increased students' performance levels. Significantly, students under game-based learning completed the task in lesser time as compared to traditional learning.



Wearable devices have impacted the daily life of every individual. These devices come with the embedded feature that fits almost within clothes, accessories, or even watches. One of the wearables named "wrist-worn devices" has gained acceptance by the masses among other wearables. Vital information can be easily gathered with the help of such wearables. These are also suitable for myriad applications such as sports, agriculture, medical, and several more. This chapter gives a comprehensive review of wearable computing electronic devices being used in various fields and provides the latest trends in wearables. The main objective of this review is to discuss various challenges that are faced by individuals in using wearables and the latest methods that can overcome these issues. In the modern epoch, wireless wearable devices have gained a lot of importance in various fields such as sports, agriculture, medical, and many more.



The electroencephalogram is used in brain-computer interface (BCI) in which signal from the human brain is sensed with the help of EEG and then sent to the computer to control the external device without having any touch of muscular body parts. On the other hand, the brain chip interfacing (BCHIs) is a microelectronic chip that has physical connections with the neurons for the transfer of information. The BCI needs a reliable, high-speed network and new security tool that can assist BCI technology. 5G network and blockchain technology is ideal to support the growing needs of brain chip interfacing. Further, the Cloudmind, which is an emerging application of BCI, can be conceptualized by using blockchain technology. In this chapter, brain-computer interfaces (BCIs) are expedient to bridge the connectivity chasm between human and machine (computer) systems via 5G technologies, which offers minimal latency, faster speeds, and stronger bandwidth connectivity with strong cryptographic qualities of blockchain technologies.



Neuroscience is the study of the brain and its impact on behavior and cognitive functions. Computational neuroscience is the subfield that deals with the study of the ability of the brain to think and compute. It also analyzes various electrical and chemical signals that take place in the brain to represent and process the information. In this chapter, a special focus will be given on the processing of signals by the brain to solve the problems. In the second section of the chapter, the role of graph theory is discussed to analyze the pattern of neurons. Graph-based analysis reveals meaningful information about the topological architecture of human brain networks. The graph-based analysis also discloses the networks in which most nodes are not neighbors of each other but can be reached from every other node by a small number of steps. In the end, it is concluded that by using the various operations of graph theory, the vertex centrality, betweenness, etc. can be computed to identify the dominant neurons for solving different types of computational problems.



To understand more about the human brain and how it works, it is vital to understand how the neural circuits connect different regions of the brain. The human brain is filled predominantly with water and the majority of the water molecules undergo diffusion which can be captured with the help of diffusion MRI. Diffusion weighted images enable us to reconstruct the neural circuits in a non-invasive manner, and this procedure is referred to as tractography. Tractography aids neurosurgeons to understand the neural connectivity of the patient. This chapter attempts to explain the procedure of tractography and different types of algorithms.



Cerebral venous thrombosis (CVT) is a rare condition involving various symptoms that is mainly seen in younger adults. The most commonly involved are the superior

The transient ischemic attack is a neurological emergency which is a clinical view of focal cerebral, retinal, or spinal dysfunction that lasts less than an hour, without any detectable acute infarction in neurological imaging methods. TIA is a serious warning for ischemic stroke, and this risk is particularly high in the first 48 hours. Following TIAs, approximately 10-15% of patients undergo stroke in 90 days and about half of these patients suffer a stroke in the first two days. Neuroimaging and laboratory studies should be performed quickly to reveal the etiology and to reduce the risk of stroke that may develop in patients present with TIA. Therapeutic and preventive interventions should be started as soon as possible. With early diagnosis and treatment, the risk of a 90-day stroke in these patients can be reduced by 80%. In addition to antiplatelet and anticoagulant treatments, aggressive control of blood pressure, regulation of blood sugar, statin, dietary recommendations, exercise, and managing the other underlying specific conditions should be started quickly.



The trunk is the part of the human body that provides basic mechanical stabilization. It provides strength transmission between the upper and lower body regions. Body control is the ability of the body muscles to maintain the upright posture, to adapt to weight transfers, and to maintain selective trunk and limb movements by maintaining the support surface in static and dynamic postural adjustments. Good proximal trunk control provides better distal limb movements, balance, and functional motion. There are many evaluation methods, devices, and scales for trunk function and performance. 3D kinematic, electromyography, hand-held dynamometer, isokinetic dynamometer, trunk accelerometer are some devices that measure trunk function. The motor assessment scale-trunk subscale, the stroke impairment assessment set-trunk control subscale, trunk control test, trunk impairment scale are the most used scales. This chapter explores the effect of strokes on the trunk.

Vestibular rehabilitation (VR) is a therapeutic approach prepared specifically for each individual who has a vestibular and balance disorder. VR helps in the treatment of unilateral or bilateral vestibular hypofunction and vestibular problems such as labyrinthitis and vestibular neuronitis. Individuals who have inner ear problems which have not been solved for a long time or have received medical treatment benefit from VR. In addition, VR helps to alleviate the complaints of individuals who have undergone surgery due to vestibular problems. With the VR program, regulative activities are carried out to decrease the duration, intensity, and frequency of vertigo; the symptoms of vertigo; increase independency in daily life activities; and to make it possible for patients to deal with the feelings of dizziness, imbalance, and anxiety, in addition to training patients about this issue and regulating the general conditions. The aim is to increase the life quality of patients.

# Preface

Nearly each day novel neuroscientific findings show that the researchers are focusing on developing advanced smart hardware designs and intelligent computational models to imitate the brain's computational abilities. Their key focus is on theoretical analysis and mathematical modeling of neurons & brain systems to mimic the functional behavior of the nervous system at circuit, biophysical and deep system levels and repairing artificial damaged systems in the human body. Neuroengineering elaborates the contemporary and futuristic technological designs and implementation techniques to interact with the human brain by integrating mathematical, biological, chemical & engineering tools.

The advancements in smart computational techniques and materials provide a significant role in inventing intelligent bioelectronic device designs with smart features such as accuracy, low power consumption etc. without damaging the human tissues. These invasive & non-invasive based smart bioelectronic devices made easy & better understanding of internal functions of the human brain.

The data acquired with bioelectronic smart devices is massive and sensitive. For management & analysis of this crucial data, the use of advanced computational and learning algorithms plays a significant role in neuroengineering. These advanced and intelligent computing models through machine and smart deep learning algorithms help to understand the information processing capabilities of the human brain with optimum accuracy.

The collaboration and contribution of lead researchers from academia, industry and clinical medicine, neuroengineering becomes successful in designing advanced contemporary & futuristic designs and models in neuroscience which helps in diagnosis and understanding the physiology of the subject and also in developing optimum technological intervention through interdisciplinary approach.

The technological advancements made it easy for the researchers to understand the new horizons in field of diagnosis, analysis and treatment for various neurological diseases. The readers/researchers should understand the clinical features of neurological disorders through various advanced computational algorithms such as machine learning, deep learning etc. in advanced communication systems such as 4G, 5G and beyond.

This book includes the fundamental computational and modeling techniques in neuroscience and also the contemporary & futuristic computational and neuroengineering designs and algorithms for better understanding of neurological disorders. The prominence topics this book will include brain machine interface, collection of sensitive data (invasive or non-invasive), generating neurological databases, advanced computational neuroscience models for the diagnosis of disorders, clinical features of various neurological disorders and providing technological interventions to make a healthy lifestyle.

## INSIDE THIS BOOK

Chapter 1 empathizes on comparative analysis of ANN such as Pattern Recognition Neural Network (PRNN), Feed Forward Neural Network (FFNN) and Support Vector Machine (SVM) based classifiers for human machine interface based wheel chair control followed by feature extraction through Discrete Wavelet Transform (DWT).

Chapter 2 focuses on investigation of the impact of game-based learning on students' performance. An android-based game intervention named "KidsZoneApp" was developed which included Mathematics and English lesson plans according to the student's curriculum. GSR NUL-217 logger sensor was used to record the physiological signals of each participant in real-time. It was observed that game-based learning contributed to increased students' performance levels.

Chapter 3 highlights the wearable computing electronic devices being used in various fields and provides the latest trends in wearables. Now a days, wireless wearable devices have gained huge importance in various fields such as sports, agriculture, medical and many more. Indeed, these wearable devices positively and progressively impacted the daily life of every individual.

Chapter 4 highlights that Brain-Computer Interfaces (BCIs) are expedient to bridge the connectivity chasm between human and machine (computer) systems via 5G technologies, which offers minimal latency, faster speeds and stronger bandwidth connectivity with strong cryptographic qualities of block chain technologies.

Chapter 5 focuses on the importance of graph theory for computational neuroscience. Graph-based analysis reveals meaningful information about the topological architecture of human brain networks. It was observed in the chapter that graph theory can be utilized to identify the dominant n eurons for solving different types of computational problems.

Chapter 6 focuses in procedural structure of tractography and different types of algorithms. Human brain is filled predominantly with water and majority of the water molecules undergo diffusion which can be captured with the help of diffusion MRI. Diffusion weighted images enables us to reconstruct the neural circuits in a non-invasive manner and this procedure is referred to as tractography.

Chapter 7 focuses on rare Cerebral Venous Thrombosis (CVT) condition observed in younger adults. Neurological signs including motor and sensorial losses, impaired consciousness, speech disorder, epileptic seizures, visual problems (hemianopia, nystagmus, diplopia, and papilledema), and cranial nerve signs may be seen. The diagnosis can be made primarily by suspecting the clinical condition and radiological presentation of thrombosis.

Chapter 8 discusses the clinical features of stroke which causes neurological dysfunction due to focal infarction or haemorrhage in the brain, spinal cord or retina. Age, gender, race, heredity, ethnicity, hypertension, atrial fibrillation, diabetes, hyperlipidemia, smoking, transient ischemic attack and physical inactivity are risk factors of stroke. Mental dysfunction, speech and language disorders, motor and sensory impairment may occur as a result of stroke.

Chapter 9 focuses on the complications of strokes and symptoms & signs, diagnosis and management of these comorbidities and complications. It highlights diagnosis and treatment of cardiac problems, sleep disorders, deep vein thrombosis, pulmonary embolism, dysphagia, malnutrition and pneumonia. Depression, central post-stroke pain, upper limb problems after stroke, spasticity, bladder dysfunction are also discussed in this article.

Chapter 10 highlights that Transient Ischemic Attack (TIA) is a neurological emergency which is a clinical view of focal cerebral, retinal, or spinal dysfunction that lasts less than an hour, without detectable any acute infarction in neurological imaging methods. The results shows that with early diagnosis and treatment, the risk of a 90-day stroke in these patients can be reduced by 80%. In addition to antiplatelet and anticoagulant treatments, aggressive control of blood pressure, regulation of blood sugar, statin, dietary recommendations, exercise, and managing the other underlying specific conditions should be started quickly

Chapter 11 highlights that trunk provides basic mechanical stabilization and strength transmission between the upper and lower body regions in humans. Good proximal trunk control provides better distal limb movements, balance, and functional motion. This chapter also highlights some evaluations methods and trunk impairment scales.

Chapter 12 focuses vestibular and balance disorder and also explains the anatomy & physiology of vestibular system. Vestibular rehabilitation (VR) helps in the treatment of unilateral or bilateral vestibular hypofunction and vestibular problems such as labyrinthitis and vestibular neuronitis, which leads to increase in life quality of patients.

While shaping the book, this was kept in mind that the technological and clinical issues regarding computational neuroscience should be focused. The first section from chapter 1 to chapter 7 provides technological improvements to detect, analyze and diagnose the neurological disorders. The rest of the chapters describes the clinical view of neurological disorders and their possible remedies.

A very warm thanks to the very hard working to the excellent contributors/authors who made tremendous efforts to brief the findings and possibilities in chapters. This book will be a source for knowledge in the field of computational neuroscience and for students/researcher and readers, this book will provide complete bridge between technological and clinical features and improvement in the field of computational neuroscience.

In future, we encourage the contributors for this book to collaborate with international community to enhance their work.

*Harjit Pal Singh*
*Anurag Sharma*
*Vikas Khullar*

# Chapter 1
# Human-Machine Interface-Based Robotic Wheel Chair Control

**Deepti Kakkar**
https://orcid.org/0000-0002-9681-1291
*Dr. B. R. Ambedkar National Institute of Technology, India*

**Ashish Raman**
https://orcid.org/0000-0002-6720-373X
*Dr. B. R. Ambedkar National Institute of Technology, India*

## ABSTRACT

*This chapter presents the P300-based human machine interference (HMI) systems control robotic wheel chair (RWC) prototype in right, left, forward, backward, and stop positions. Four different targets letters are used to elicit the P300 waves, flickering in the low frequency region, by using oddball paradigms and displayed on a liquid crystal display (LCD) screen by Lab-VIEW. After the pre-processing and taking one second time window, feature is extracted by using discrete wavelet transform (DWT). Three different classifiers—two based on ANNs pattern recognition neural network (PRNN) and feed forward neural network (FFNN) and the and other one based on support vector machine (SVM)—are used. Those three techniques are designed and compared with the different accuracies among them.*

DOI: 10.4018/978-1-7998-7433-1.ch001

## INTRODUCTION

Human Machine Interface (HMI) system is unique of wildest rising field of investigation and growing rapidly now a day (Farwell & Donchin, 1988). HMI system is a normal form of communication, requires peripheral nerves and muscles. it is started by user intension (Wolpaw et al., 2002). This intent initiates a process in which certain brain areas are activated, and hence signal are directed via peripheral nerves system, to the corresponding muscles, which in turn executes the movements required for the communication or controlling task (Wolpaw & McFarland, 2003). HMI provide an another way to the natural communication as well as it helps disabled persons to control their activities (Mason & Birch, 2003). Generally two ways: invasive and non- intrusive HMI are commonly used for developing natural communication.

**Invasive HMI:** Obtrusive BCI gadgets are inserted inside the skull however rest outside the cerebrum as opposed to inside the gray matter (Zhu et al., 2010). Signal quality utilizing this sort of BCI is bit smaller when it looks at to non-obtrusive BCI. They deliver preferred determination signals over non-intrusive BCI. In partially invasive BCIs have less danger of scar tissue development when contrasted with invasive BCI (Zhu et al., 2010). Electrocochleography [ECoG] utilizes the same innovation as non-obtrusive EEG, yet the anodes are inserted in a dainty plastic pad that is set over the cortex, underneath the dura mater (Towle et al., 1993). ECoG advancements were first exchange human in a later trial; the scientists empowered a high school kid to play space intruders utilizing his ECoG insert (Towle et al., 1993). This examination shows that it is hard to create kinematics BCI gadgets with more than one measurement of control utilizing ECoG (Towle et al., 1993). Light responsive imaging BCI devices are still in the area of hypothesis. These would include embedding laser inside the skull. The laser would be prepared on a solitary neuron and the neuron's reflectance measured by a different sensor. At the point when neuron fires, the laser light example furthermore, wavelengths it reflects would change marginally (Towle et al., 1993).

**Non-Intrusive HMI:** Non- intrusive BCI has the minimum sign clarity with regards to speaking with the cerebrum but it is thought to be most secure at the point when contrasted with different sorts. This kind of gadget has been observed to be effective in giving a patient the capacity to move muscle embeds and reestablish fractional development (Nicolas-Alonso & Gomez-Gil, 2012). A non-obtrusive procedure is one in which therapeutic checking gadgets or sensors are mounted on tops or headbands read cerebrum signals. This methodology is less meddlesome additionally perused flags less viably in light of the fact that cathodes can't be set specifically on the sought part of the mind. A standout amongst the most well-known gadgets under this classification is the EEG equipped for giving a fine fleeting determination. It is anything but difficult to utilize, shabby and portable (Nicolas-Alonso & Gomez-Gil, 2012).

HMI is an un-natural system that by passes body usual affects paths which are the neuromuscular output channels (Cheng et al., 2002). Instead of depending up on peripheral nerves and muscles, HMI straightly processes mind actions associated to the users intensions and translates the brain actions which are recorded, into respective control signals for HMI applications (Erdoğan, 2009). A computer is used to translate signal processing and pattern recognition because the measured actions directly emerged from the brain not from peripheral nerves, this system is called as Human Computer Interface [HCI]. Some of the brain activities that can be successfully noted from the scalp by using EEG are Event Related Potentials (ERPs), Slow Cortical Potentials (SCPs), P300 potentials and Steady-State Visual Evoked Potentials (SSVEPs) (Thulasidas et al., 2006; Zhu et al., 2010). Among them P300 are attracted due to its advantages of requiring less or no training, high information training rate (Towle et al., 1993).

Some of the main aspects that can regulate the performance of a Brain Computer Interface [BCI] system include the type of the brain signal used to transfer the intentions, feature extraction methods, classification algorithms to achieve the control commands etc (Nicolas-Alonso & Gomez-Gil, 2012; Rakotomamonjy & Gigues, 2008)., the proposed work explain the effect of three different classification methods in improving the performance of a P300 based wheel chair control system. Proposed work can govern a wheel chair in forward, right, left, backward and in stop positions. Among these classifiers, two based on Artificial Neural Network (ANN) and one based on Support Vector Machine (SVM) are compared with each other (Garrett et al., 2003).

This chapter aimed to detect the P300 wave accurately. Brain-controlled BCI system follows some steps such as signal acquisition, feature extraction, feature selections, classifications and hardware interfacing. P300 signals, flickering in the low-frequency region, by using oddball paradigms and displayed on a Liquid Crystal Display (LCD) screen by Lab View. Paradigms contain 36 alphanumeric signs and placed into 6 by 6 matrices. EEG signals are recorded from ten subjects of aged between 22-24 years. After pre-processing and taking one second time window, signals are ready for feature extraction. Discrete Wavelet Transform (DWT) technique is used for feature extraction 5$^{th}$ level of decompositions is done by using db4 and obtained detailed approximated coefficients are used as a feature.

## Proposed Methodology

*Figure 1. The proposed method RWC prototype control by using P300 HMI application.*

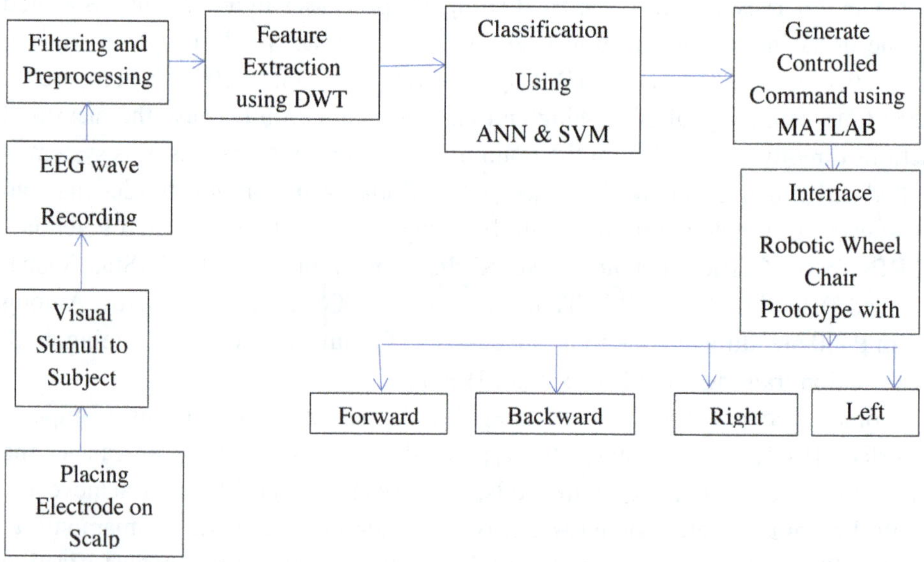


## 1. Subject

A total of ten healthy subjects are participating in this experiment of age between 22 -24 years, all are male. All are the right handed and normal vision. None of the participant has participated previously in visual P300 BCI study experience.

## 2. Visual Stimuli Paradigm

The visual stimuli paradigm for generating a P300 wave can be presented on PC screen (Towle et al., 1993). This is programmed by using Lab-View software, composed by 36 different letters including alphabets and numerical digits with colored in block background. Instead of flashing each letter, complete row or column flashed in random manner and color of flashed letter becomes red during flash (Nicolas-Alonso & Gomez-Gil, 2012). Row and column flashed for 100ms and blank screen appear for 200ms thus time period between each flash is 300ms. Row or column flash for 12 times that is called as one repetition and total 10 repetitions is set and inter- repetition time delay is 1 sec. Group of 10 repetitions is said trail and total 5 trail were set and inter trial time delay is 5 sec. as shown in Figure 2

*Figure 2. Proposed visual stimuli paradigm of P300 detection.*

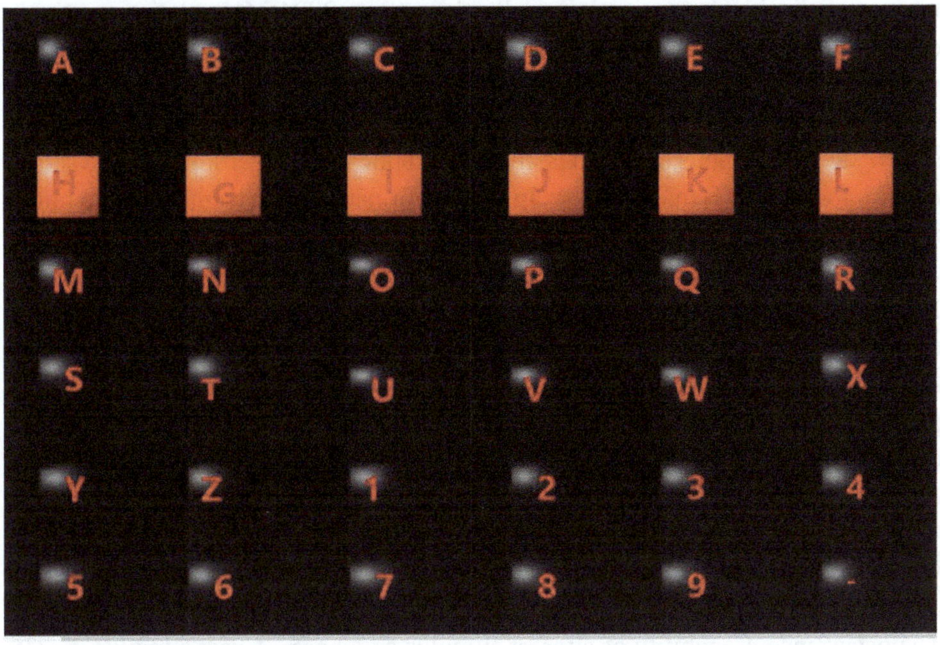


## 3. EEG Recording

Essentially, an EEG is utilized to quantify the mind action from the scalp as electrical signal (Nicolas-Alonso & Gomez-Gil, 2012). An EEG is characterized as the electrical development of a pivoting sort recorded from the scalp surface consequent to being gotten by brain anodes, conductive broadcasting (Dean, 2008). The EEG sign is only the present estimation among the dendrites of courage cells in the area of the mind amid their synaptic excitation (Dean, 2008). This present comprises of electrical arena measured by EEG gear and the attractive field evaluated by EMG gadgets. Human mind has 3 unique areas in particular; cerebrum, cerebellum and mind stem (Dean, 2008). Every range of the brain speaks to the diverse status of the human body like the cerebellum area speak to the start of development, awareness and perspective, the cerebellum district assumes a part in intentional activity like power associated development and the cerebrum stem locale regulator the breath working and neural hormones (Maggi et al., 2006). Soit is particularly clear to an EEG sign produced from cerebrum unquestionably definite the position of the entire physique and mind issue (Maggi et al., 2006).

**EEG Signal Recording:** In routine scalp EEG, the record is gotten from various anodes orchestrated in a specific example on the scalp with a conductive glue or gel. Anode territories then labels are shown by the international 10–20 framework for best medical and examination applications (Ben, 2004). Calculating terminals deliver the sign from the scalp which is in the extent of microvolt in this manner intensifier is utilized to increase the bio-potential into the reach everywhere they canister be digitized precisely then Analog to Digital (A/D) converter, changes over simple sign to computerized sign lastly PC or applicable gadget supplies and shows noted information (Ben, 2004). Scalp records of neuronal movement in the mind, recognized as an EEG, measures the latent variations after some time in essential electronic path leading among the sign (dynamic) and a position terminal. Additional 3rd cathode, named the ground terminal, is essential for receiving differential energy by the same energies appearing at dynamic and orientation focuses. The mono-channel EEG estimation includes single dynamic cathode, maybe a couple particularly connected composed, orientation and one ground cathode. The multi-channel courses of action might include up to 100 or 200 dynamic terminals (Kim & Sung-Phil, 2011).

There were 36 target possibilities available, out of 36 only 4 center target letters were selected to be focused by subjects, because we need only four classes for controlling a Robotic Wheel Chair prototype. These letters were O, U, V and P. The occurrence probability of a target stimulus and a non –target stimulus is 1/36 and 35/36 respectively.

The EEG signal is recorded at the $O_z$, $C_z$, $P_z$, $P_3$ and $P_4$ positions according to the 10-20 international electrode system by using RMS EEG-32super spec system (Recorder and Medicare system, India). One ground and one reference electrode was placed on forehead at position $F_{PZ}$. Along with all these electrodes two more electrodes are placed bottom tip of left and right ear that were known as earlobe which are named as $A_1$ & $A_2$ for left and right ear respectively. None of the impedance between scalp electrode and the reference electrode exceeded above the 10kΩ. EEG signals are filtered by using band pass filter with1Hz and 35Hz lower and upper cut-off frequencies respectively. Signals are sampled at 250 Hz and sensitivity kept at 7.5 μV/mm) (Ben, 2004; Blankertz et al., 2004; Cocotte & Graser, 2011; Dean, 2008; Garrett et al., 2003; Kim & Sung-Phil, 2011; Maggi et al., 2006; Rosas-Cholula, 2010). Before recording the EEG signal, subject sits at a distance of 60cm from computer screen and subject is instructed to focus at particular letter out of four selected letters and count the number of flash of the target letter because this is easy way to keep the focusing on target. Also subjects are instructed to avoid the eye blinking during recording session. Paradigm started to play for 2 minutes only because initial trial was run for training so that subject learnt to focus on target as shown in Figure 3.Total paradigm run time for one target was 4 minutes and this is called one session. The time interval between each session was 10 minutes and four sessions were recoded for 4 target classes. Thus it takes 56 minutes for one subject (Maggi et al., 2006).

*Figure 3. Block diagram of P300 wave recording experiment set up.*

## 4. Data Prepressing

After signal acquisition it is filtered using the 4th order Butterworth band pass filter of frequency range from1Hz to35Hz (Kaper et al., 2004). This filter covered significant components of P300 wave and remove unwanted signals. The notch filter was set to 50Hz depending upon the power supply. Signals obtain from the channel $C_Z$-$A_2$ for further process by its maximum SNR more probability to find P300 EPR'S.

## 5. Feature Extraction

Feature extractions are done by DWT because of its dynamicity in time and frequency. Signals are recorded from various scalp positions but out of all those positions $C_z$-$A_2$ channel are selected for further processing because of its high SNR ratio. After preprocessing and taking one second time window, obtained signal is decomposed by using DWT. Fifth level of decomposition is done by using db4 and after that reconstruction is obtained. Each decomposition is obtained at different frequency range is at 4th level detailed coefficient is in 15-35Hz frequency at 5th level detailed coefficient is in 7-15Hz range and the 5th level approximated coefficient was in 0.5-7 Hz frequency range hence a parts of the other detailed coefficient and approximated coefficient is not in the frequency range hence there was ignored (Ben, 2004). MATLAB is used for developing the DWT program. Here five detailed and one approximated signal is obtained. The graph is plotted between amplitude in μV and time in ms as shown in Figure 4. In signals $A_5$ and $D_5$ it could be seen that a negative peak is coming near to 300 ms that is our P300 peak (Kim & Sung-Phil, 2011). But out of these six signals, detailed and approximated coefficients at 5th level and also detailed coefficients at 4th level are same in range of P300 signal range so these coefficients are selected as features (Rosas-Cholula, 2010).

## 6. Feature Selection Method

This wavelet (Übeyli, 2009) coefficient is used as feature for classification. A transform is a method for remapping the first sign which gives a larger number of information than the first signal (Lin & Zhang, 2005). There are various methods for playing out this assignment in literature like Fourier Transform, Short Time Fourier Transform (STFT), Fast Fourier Transform (FFT) and so on. In any case, none of them can portray the time and frequency data of the waveform immediately which prompts the change of another change to determine above issue named as Wavelet Transform (Lin & Zhang, 2005). The expression "wavelet" insinuates an oscillatory vanishing wave with time-restricted develop, which can depict the time-frequency flat, by particles of various time underpins. All around, wavelets are deliberately made to have exact assets that mark them profitable used for indication handling. They speak to a reasonable mechanical assembly for the examination of non-stationary (Lin & Zhang, 2005).

The STFT gives the indistinguishable determination at all rate of recurrence, however the utilization of wavelet transform gave an importance preferred improvement to analyses the diverse rate of recurrence of the signs with changed recurrences by the strategy known as multi-determination system (Blankertz et al., 2002; Matthias Kaper, 2004). The wavelet transform evidences its convenience to depict the assets of the waveform that progressions with time by isolating the waveform into a portion of rule. In this manner, the wavelet change is especially appropriate for researching the EEG signs (Ioana et al., 2007; Schlögl et al., 2005).

At 4th level 24 detailed coefficient and at 5th level 16 detailed and 16 approximated coefficients was obtained. Thus total 56 features are obtained but all of these 56 features are not useful, so we have to reduce these redundant features. Now feature selection is important issues because classification accuracy depends upon feature selection. So after observation and analysis fifth level coefficients, 3 features, are selected as given below.

a. Out of the 16 fifth level detailed coefficients, 8th component is selected and name as $CD_5$-8.
b. Out of the 16 fifth level approximation coefficients, 6th component is selected and name as $CA_5$-6.
c. Out of the 24 fourth level detailed coefficients, 10th component is selected and name as $CD_4$ -10.

These selection features are varying in a particular range corresponding to each class and this range is uniformly distributed. These selected features are used as classifications.

*Figure 4. Proposed DWT at 5th level decomposition of the P300 wave.*

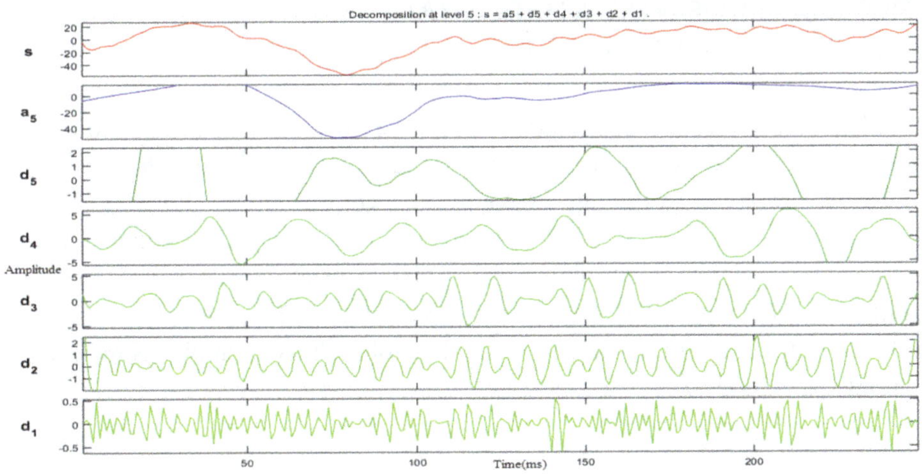


## 7. Classification

Classification is one of the most commonly encountered decisions making so here classification is done by two different methods one is SVM and other is ANN.

### a. ANN Classifier

ANN classifier technique is used in this work. Algorithm code is designed by using MATLAB software, normalized in the range of [0,1]. 'Pattern net' used for PRNN and 'Feed Forward net' used for FFNN. For 'Pattern net' scaled conjugate gradient training algorithm is used and for 'Feed Forward net' bayesian regulation training algorithm is used (Kaper et al., 2004; Lin & Zhang, 2005; Rakotomamonjy & Gigues, 2008; Thulasidas et al., 2006).

### b. 2. SVM Classifier

SVM classifier technique is also used in this work. Algorithm code is designed by using MATLAB software, normalized in the range of [0,1].Out of total sampled data 80% of samples serve as the training sets and remaining as testing set. MATLAB SVM program has to be designed for four input matrix SVM, each one for different stimuli.by using Different kernel function in SVM program result are obtained (Blankertz et al., 2004; Cocotte & Graser, 2011; Garrett et al., 2003).

## PROPOSED ROBOTIC WHEELCHAIR PROTOTYPE INTERFACING

P300 based HMI application is RWC prototype shown in Figure 5. It was designed in Arduino software environment and interfaced with HMI. The user had to control this RWC prototype by giving its controlling commands as 'Forward', 'Left', 'Right', 'Backward' and' Stop'. Prototype contains ATtmega328 Atmel company microcontroller which is Arduino Uno, 16MHz crystal oscillator, L329D motor driving IC, DC motors, FTDI 232R serial port and virtual port (Blankertz et al., 2002; Garrett et al., 2003; Ioana et al., 2007; John, 2004; Matthias Kaper, 2004; Schlögl et al., 2005).

*Figure 5. Proposed RWC prototype.*

If user wants to operate this RWC prototype then he must have to show his intention by focusing at a particular letter flashing on paradigm on computer screen. When user focused on letter 'O' then prototype will move forward. For moving it on left direction user has to focus on letter 'V'. When user focuses on letter 'U' prototype moves to right and moves in reverse direction when user focuses on letter 'P'.

Apart from these letter if user focuses on any other letter RWC prototype will stop. DC gear motors are used in this system for moving it in various directions. Each motor moves when it is given commands in two formats like 0 1 or 1 0 shown in Table1.It was connected to computer via virtual port interfaced with MATLAB software via Arduino package shown in Figure 6.

*Figure 6. Proposed RWC prototype interfacing.*

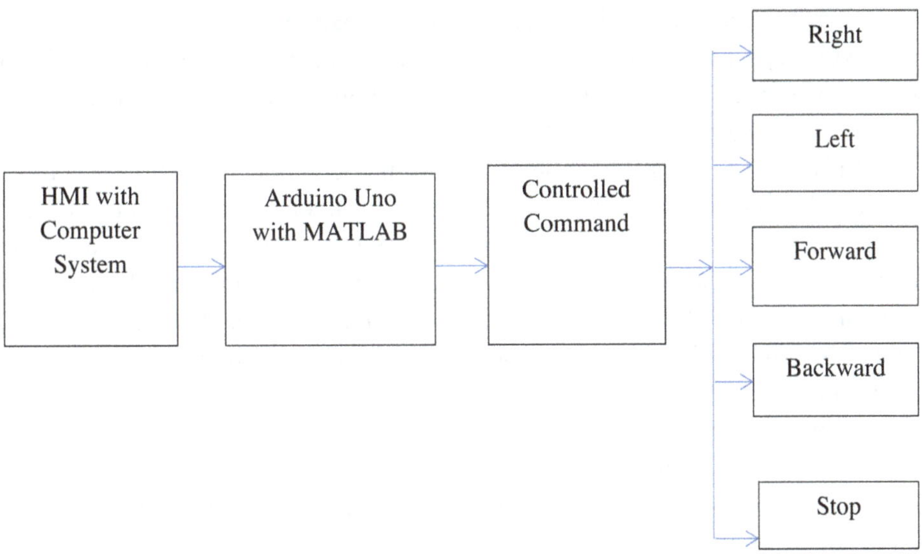


*Table 1. Designed controlled signal for RWC prototype.*

| Controlled Commands | Movements |
|---|---|
| 0 1 0 1 | Forward |
| 1 0 1 0 | Backward |
| 0 1 0 0 | Left |
| 0 0 0 1 | Right |
| 0 0 0 0 (or) 1 1 1 1 | Stop |

# RESULTS AND DISCUSSION

## ANN Classifier Results

ANN is an excellent tool for classification. Different type of network can be designed and training algorithm can also be chosen which gives different performances. The scaled conjugate gradient training algorithm for PRNN and bayesian regulation or back propagation training algorithm for FFNN are used. For both algorithms data division was same. 70% used for training, 15% used for validation and remaining 15% used for testing.

In PRNN also two input matrices are designed for this classification. One was input matrix and other one was target matrix. Number of hidden layer neuron is 10 and training algorithm is scaled conjugated gradient. Total 12 iterations run in almost zero second and 6 validations are checked. Best validation performance is obtained at 6 epochs shown in Figure 7.Overall performance of network is 0.867 and gradient is 0.689. The shown in Figure 8 and 9, regression and confusion matrix, the regression value obtained are 0.954.

In FFNN, for this classification two input matrices were design. One is input matrix and other one is target matrix. Number of hidden layer neuron is 10 and training algorithm is Bayesian regulation. Total 199 iterations run in 11 seconds and no validations are checked. Best training performance is obtained at 199 epochs as shown in Figure 10. Overall performance of network is 0.933 and gradient is 0.911. Figure11- 12 shows the regression plot and confusion matrix of FFNN. Regression value is obtained 0.984.

The confusion matrixes used for all four classes. This is one way of Showing performance which contained row and column in which row belongs output class and column belongs to target classes. Diagonal element of matrix show how much percent of data mismatched or wrongly recognized.

*Figure 7. Best validation performance of PRNN.*

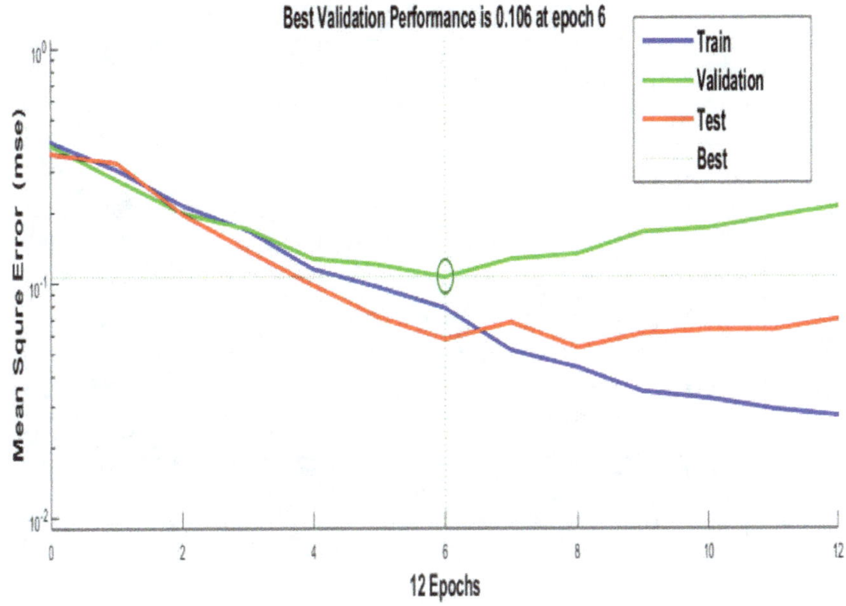


*Figure 8. Regression plot of PRNN.*

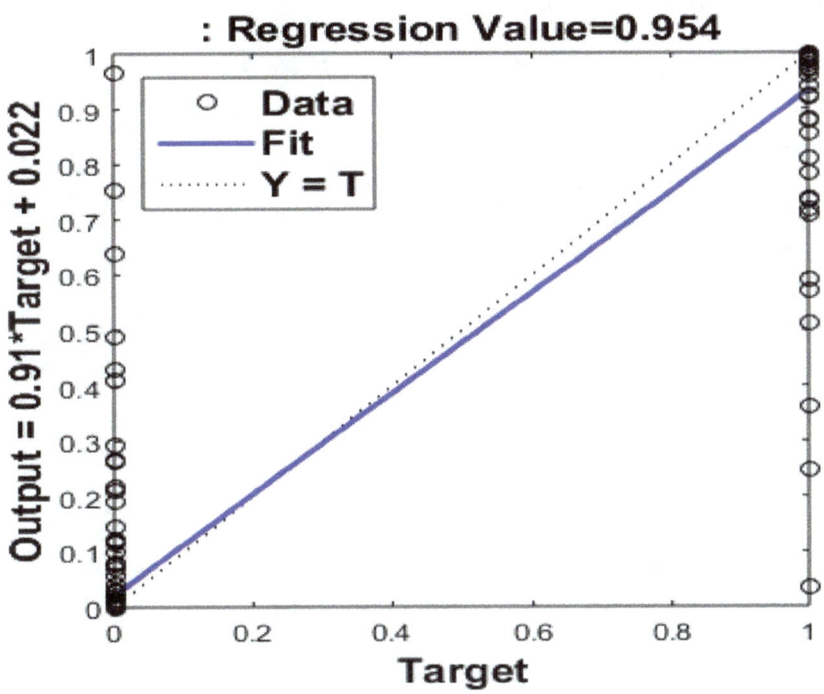

*Figure 9. Confusion matrix of PRNN.*

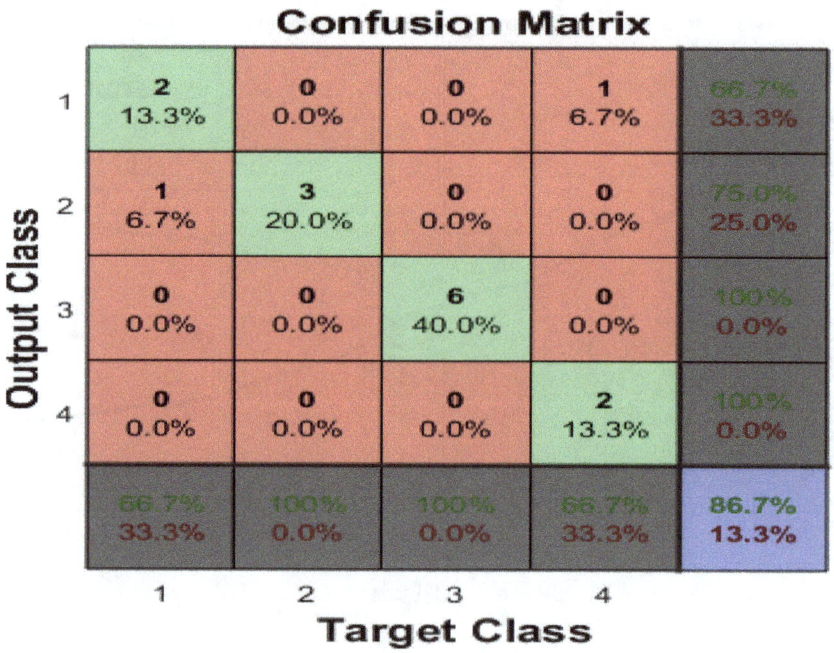


*Figure 10. Best training performance of FFNN.*

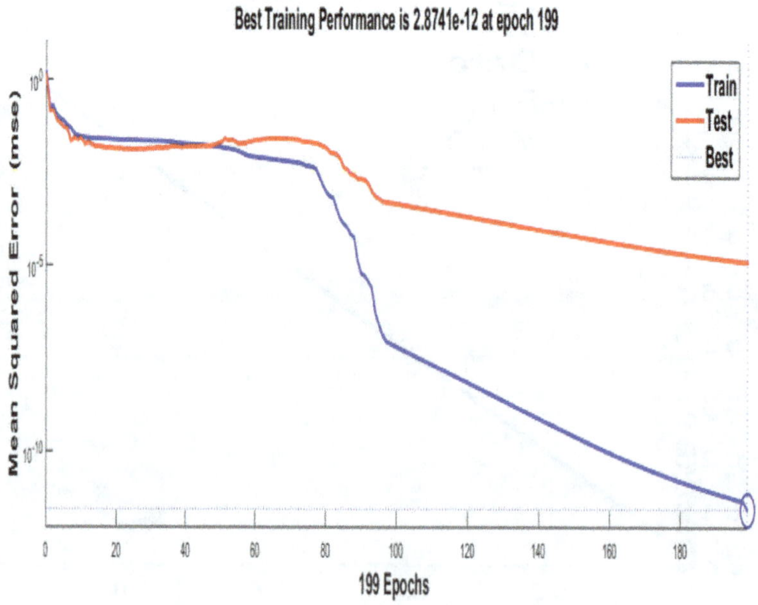

*Figure 11. Regression plot of FFNN.*

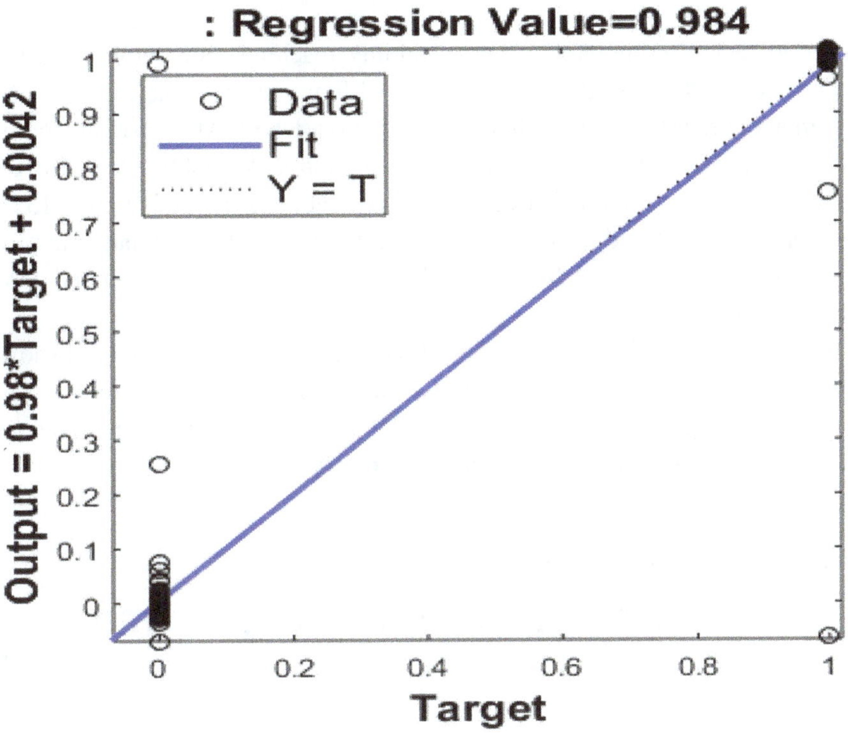


*Figure 12. Confusion matrix of FFNN*

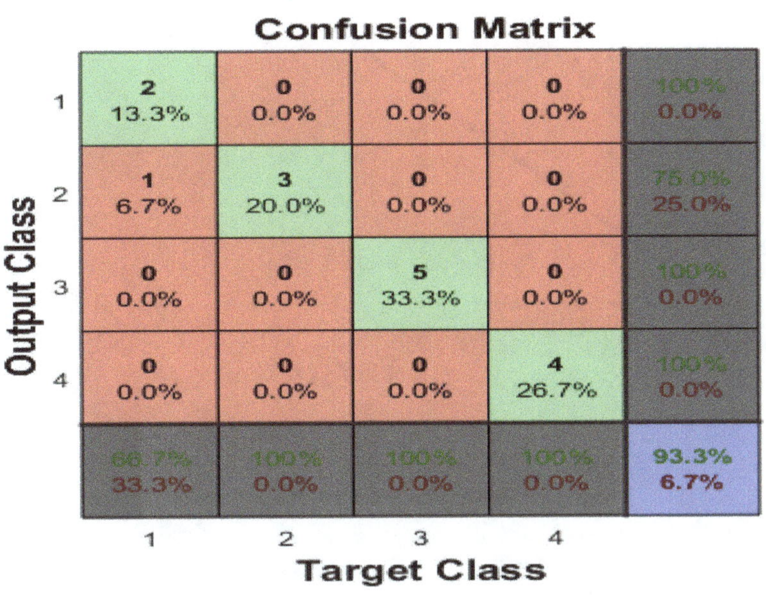

## SVM Classifier Result

This test data was completely new to training machine. For obtaining a better performance SVM is tested on different kernel functions such as linear, Gaussian polynomial of order fifth and Radial Basis Function (RBF). After so many trainings and tastings, best performance is obtained at polynomial kernel function with order fifth. Regression plot obtained after SVM classification is shown in Figure 13. Best regression value obtained at fifth order kernel function is 0.9999 that is improved accuracy and it is close to 1.0.and overall performance value is obtained 1.00 as shown in Figure 14.

The confusion matrixes are used for all four classes. This is one way of showing performance which contained row and column in which row belongs output classes and column belongs to target classes. Diagonal element of matrix shows how much percent of data mismatched or wrongly recognized.

*Figure 13. Regression plot of SVM.*

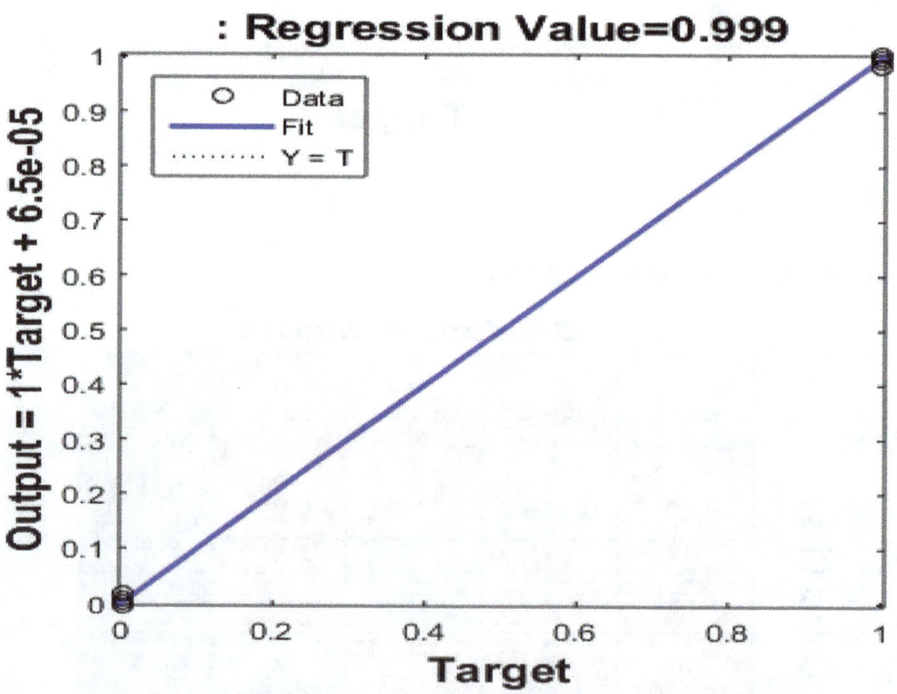

*Figure 14. Confusion matrix of SVM.*

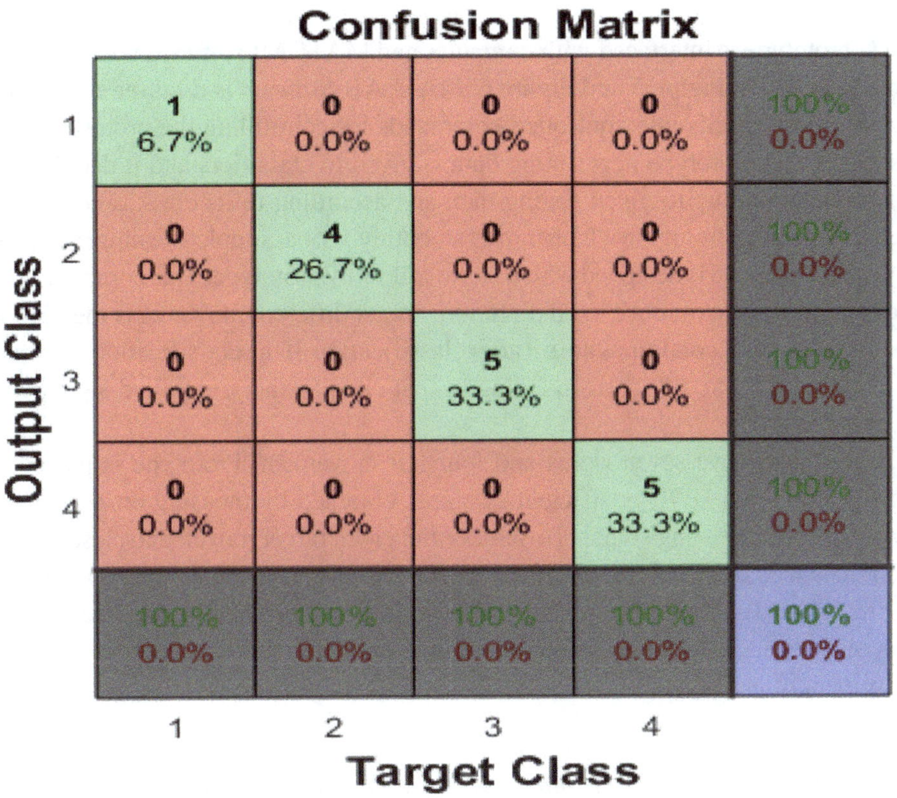


*Figure 15. Proposed path designs for RWC prototype.*

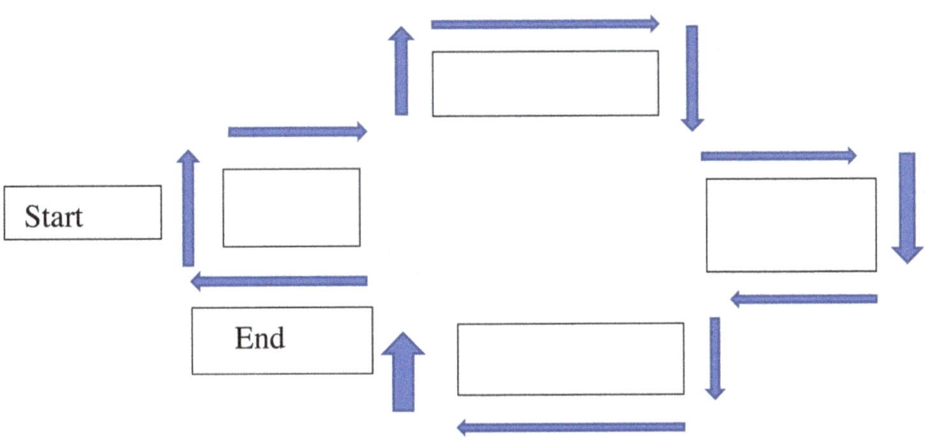

## Robotic Wheelchair Prototype Interfacing Results

RWC prototype is interfaced with computer and MATLAB software with Arduino Uno as shown in Figure 5 and Figure 6. MATLAB software is used for converting classification results into controlled commands for controlling the prototype. For testing of this prototype new testing data is given to classifiers and it decides the class corresponding to given testing data and according to its class, a controlled signal is given to prototype which moves smoothly . For a complete testing, a path as shown in Figure 15 is designed which has to follow by prototype. This complete path provides prototype to move in all direction. To get different movements mentioned above, 'if' & 'else' conditions are set after classification. If the classification accuracy is in between 75 to 100 corresponding to that class further controlled signal will be given to RWC.

Given path have seven rights and four left movement. Prototype is placed at start position and DWT coefficient of signal obtained by focusing on paradigms letter is given as testing signal. To achieve the given target movements, a sequence of signals which are obtained by focusing on letters on computer screen are given to prototype. To stop the prototype we have to give any signal out of 32 targets. It is also tested in backward or reverse direction and great accuracy has found in movement of prototype.

*Table 2. Comparison of test result of each class and each subject with*

| subjects | class-1(O) | class-2(U) | class-3(V) | class-4(P) |
|---|---|---|---|---|
| 1 | 90.38 | 85.74 | 93.27 | 93.89 |
| 2 | 90.37 | 94.82 | 93.43 | 94.42 |
| 3 | 86.68 | 92.28 | 98.82 | 93.8 |
| 4 | 93.21 | 94.19 | 95.37 | 97.31 |
| 5 | 92.16 | 96.88 | 95.42 | 96.24 |
| 6 | 87.39 | 93.89 | 97.35 | 93.92 |
| 7 | 91.35 | 95.25 | 95.82 | 96.2 |
| 8 | 92.63 | 97.38 | 98.29 | 97.21 |
| 9 | 91.43 | 97.59 | 99.999 | 98.23 |
| 10 | 93.29 | 93.64 | 95.25 | 94.46 |
| **AVG** | **95.577** | **94.657** | **96.301** | **95.577** |

SVM Classifier.

As compared the test results for different kernel functions different regression values are obtained for SVM. Best regression value is obtained at fifth level of kernel function. This regression value and the overall performance are better than that of the values obtained in PRNN and FFNN techniques as shown in Table 3.Table 2 shows the comparison of test results for each class and each subject with SVM classifier. Maximum accuracy of 100 and minimum of 85.74 is obtained.

*Table 3. Comparison of test results accuracy with different classifier.*

| classifier performance | regression values | | | overall performance |
|---|---|---|---|---|
| SVM | poly5<br>0. 999 | poly3<br>0.942 | rbf<br>0.866 | 1.00 |
| PRNN | 0.954 | | | 0.867 |
| FFNN | 0.984 | | | 0.933 |

## CONCLUSION

It is become difficult to obtained low skin-electrode impedance, so from next attempt, instruct the subject to clean their hair before experiment. After acquiring of so many subjects a conclusion appears that the subjects use light dense hair oil in their daily life. Focus of subject on letter is important for generating P300. If concentration is less then it becomes difficult to detect P300 wave. 10 minutes is allowed to the subjects to have an intermitted break between the trials that minimize the visual fatigue. Electrodes are placed on 5 positions on scalps but after analysis it is came to known that at positions $C_z$ and $P_z$ positions signals have better SNR and they can be used for further processing. But in this research, signals acquired from $C_z$-$A_2$ are used to be better SNR than $P_z$-$A_2$.For feature extraction DWT is used because it reduce the data vector size without losing any significant information and also it gives the feature in time domain and frequency domain. Coefficient of these decomposed signals can be used as feature selection method. Classification is done by PRNN, FFNN and SVM and obtained the regression value as 0.954, 0.984 and 0.999 respectively. The performance of PRNN FFBPNN and SVM were 0.925, 0.867 and 0.902 respectively. SVM gives better performance. With the result obtained from the classifier we are controlling robotic wheel chair in real time.

## REFERENCES

Ben, H. (2004, June). An Exploratory Study of Factors Affecting Single Trial P300 Detection. *IEEE Transactions on Biomedical Engineering, 51*(6), 975–978. doi:10.1109/TBME.2004.826684 PMID:15188867

Blankertz, B., Curio, G., & Müller, K. R. (2002). Classifying Single Trial EEG: Towards BCI. *Advances in Neural Information Processing Systems, 14*, 157–164.

Blankertz, B., Müller, K. R., Curio, G., Vaughan, T. M., Schalk, G., Wolpaw, J. R., Schlögl, A., Neuper, C., Pfurtscheller, G., Hinter Berger, T., Schroder, M., & Birbaumer, N. (2004, June). The BCI Competition 2003: Progress and Perspectives in Detection and Discrimination of EEG Single Trials. *IEEE Transactions on Biomedical Engineering, 51*(6), 1044–1051. doi:10.1109/TBME.2004.826692 PMID:15188876

Cheng, M., Gao, X. R., Gao, S. K., & Xu, D. (2002, March). Design and Implementation of a BCI with High Transfer Rates. *IEEE Transactions on Biomedical Engineering, 49*, 1181–1186. doi:10.1109/TBME.2002.803536 PMID:12374343

Cocotte, H., & Graser, A. (2011, March). Convolutional Neural Networks for P300 Detection with Application to BCI,". *IEEE Transactions on Communications, 33*(3), 433–445.

Dean, J. (2008, January). Toward Enhanced P300 Speller Performance. *Journal of Neuroscience Methods, 167*(1), 15–21. doi:10.1016/j.jneumeth.2007.07.017 PMID:17822777

Erdoğan. (2009). *A Design and Implementation of P300 Based BCI* (Ph.D. Thesis). The Graduate School of Natural and Applied Sciences, Middle East Technical University.

Farwell, L. A., & Donchin, E. (1988, December). Talking off the Top of Your Head: Toward a Mental Prosthesis Utilizing Event-Related Brain Potentials. *Electroencephalography and Clinical Neurophysiology, 70*(6), 510–523. doi:10.1016/0013-4694(88)90149-6 PMID:2461285

Garrett, D., Peterson, D. A., Anderson, C. W., & Thaut, M. H. (2003, June). Comparison of Linear, Nonlinear, and Feature Selection Methods for EEG Signal Classification. *IEEE Transactions on Neural Systems and Rehabilitation Engineering, 11*(2), 141–144. doi:10.1109/TNSRE.2003.814441 PMID:12899257

Ioana, Nafornita, Boucher, & Isar. (2007). A Bayesian Approach of Hyper Analytic Wavelet Transform Based Denoising. *International Symposium on IEEE, Intelligent Signal Processing*, 1-6.

John, L. (2004, May). *Semmlow, Biosignal, and Biomedical Image Processing: MATLAB-based Applications*. Academic Press.

Kaper, M., Meinicke, P., Grossekathoefer, U., Lingner, T., & Ritter, H. (2004, June). BCI Competition 2003-Dataset IIb: SVM for the P300 Speller Paradigm. *IEEE Transactions on Biomedical Engineering*, *51*(6), 1073–1076. doi:10.1109/TBME.2004.826698 PMID:15188881

Kim & Sung-Phil. (2011). Detection of P300 Components using the Wiener Filter for BCI-Based Spellers. *8th Asian IEEE Conference on Control (ASCC)*, 892-896.

Lin, Z. L., & Zhang, C. S. (2005). Enhancing Classification by Perceptual Characteristics for the P300 Speller Paradigm. *Proceedings of the 2nd International IEEE EMBS Conference on Neural Engineering*. 10.1109/CNE.2005.1419688

Maggi, L., Parini, S., Puccini, L., Panini, G., & Andreoni, G. (2006). A Four Command BCI System Based on the SSVEP Protocol. *IEEE International Conference on Engineering in Medicine and Biology Society*, 1264-1267. 10.1109/IEMBS.2006.260353

Mason, S. G., & Birch, G. E. (2003, March). A General Framework for BCI Design. *IEEE Transactions on Neural Systems and Rehabilitation Engineering*, *11*, 70–85. doi:10.1109/TNSRE.2003.810426 PMID:12797728

Matthias Kaper, M. (2004, June). BCI Competition 2003-Data Set IIb: SVM for the P300 Speller Paradigm. *IEEE Transactions on Biomedical Engineering*, *51*(6), 1073–1076.

Nicolas-Alonso, L. F., & Gomez-Gil, J. (2012, January). Brain Computer Interfaces, a Review. *Sensors*, *12*(2), 1211–1279. doi:10.3390120201211 PMID:22438708

Rakotomamonjy, A., & Gigues, V. (2008, March). BCI competition III: Data Set II-ensemble of SVMs for BCI P300 Speller. *IEEE Transactions on Biomedical Engineering*, *55*(3), 1147–1154. doi:10.1109/TBME.2008.915728 PMID:18334407

Rosas-Cholula. (2010). On Signal P300 Detection for BCI Applications Based on Wavelet Analysis and ICA Preprocessing. *IEEE Conference on Electronics, Robotics and Automotive Mechanics Conference (CERMA)*, *23*, 360-365.

Schlögl, A., Lee, F., Bischof, H., & Pfurtscheller, G. (2005, June). Characterization of Four-Class Motor Imagery EEG Data for the BCI-competition 2005. *Journal of Neural Engineering*, *2*(4), 14–22. doi:10.1088/1741-2560/2/4/L02 PMID:16317224

Thulasidas, M., Guan, C., & Wu, J. (2006, March). Robust Classification of EEG Signals for BCI. *IEEE Transactions on Neural Systems and Rehabilitation Engineering*, *14*(1), 24–29. doi:10.1109/TNSRE.2005.862695 PMID:16562628

Towle, L., Bolaños, J., Suarez, D., Tan, K., Grzeszczuk, R., Levin, D. N., Cakmur, R., Frank, S. A., & Spire, J.-P. (1993, January). The Spatial Location of EEG Electrodes: Locating the Best-fitting Sphere Relative to Cortical Anatomy. *Electroencephalography and Clinical Neurophysiology*, *86*(1), 1–6. doi:10.1016/0013-4694(93)90061-Y PMID:7678386

Übeyli, E. D. (2009, March). Combined Neural Network Model Employing Wavelet Coefficients for EEG Signals Classification. *Digital Signal Processing*, *19*(2), 297–308. doi:10.1016/j.dsp.2008.07.004

Wolpaw, Birbaumer, McFarland, Furtscheller, & Vaughan. (2002). BCI for Communication and Control. *Clinical Neurophysiology*, *113*, 767-791.

Wolpaw, J. R., & McFarland, D. J. (2003, June). The Wadsworth Center BCI Research and Development Program. *IEEE Transactions on Neural Systems and Rehabilitation Engineering*, *11*, 1–4. doi:10.1109/TNSRE.2003.814442

Zhu, D. H., Berger, J., Molina, G. G., & Arts, R. M. (2010, March). A Survey of Stimulation Methods used in SSVEP-Based BCI system. *Computational Intelligence and Neuroscience*, *52*, 702–357. PMID:20224799

# Chapter 2
# Investigating the Impact of Gamification in Children Using GSR

**Sonali Rani Sahansra**
*CT Group of Institutions, India*

**Anshu Sharma**
*CT Institute of Technology and Research, India*

## ABSTRACT

*This study aimed to investigate the impact of game-based learning on student performance. For this reason, an empirical study was conducted which comprises the comparison of traditional learning and game-based learning. The participants were lower primary school students of age group 6-8 years. GSR NUL-217 logger sensor was used to record the physiological signals of each participant in real time. An Android-based game intervention named "KidsZoneApp" was developed which included mathematics and English lesson plans according to the student's curriculum. The collected log data was used to calculate the changes in different dimensions of completing the task. ANOVA yielded a very significant difference between game-based learning and traditional learning groups. Overall, the results showed that game-based learning contributed to increased students' performance levels. Significantly, students under game-based learning completed the task in lesser time as compared to traditional learning.*

DOI: 10.4018/978-1-7998-7433-1.ch002

## INTRODUCTION TO GAMIFICATION

Gaming has become an important part of people's life. Playing games is considered as a popular activity throughout the life span (Huizenga et al., 2019). People play games to release their tension. By playing games, children learn to explore and to bond with their parents and peers. Additionally, the types of games people favor to play also controls their personal development (Pandey, 2021). For example, adventurous games can help people to practice the skills or to act in the roles of who they want to be in existence, memory games can brush up their mental skills and word games can improve their communication skills (Agrafioti et al., 2012).

The broad utilization of new advances, for example, the Internet, cell phones, and social networks has influenced the learning cycle in schools and colleges (Arroyo-Gómez et al., 2017). Innovation vitally affects learning, making it conceivable to find out about new data frameworks without any problem (Ashok Charan, n.d.). One of the most common education systems, which are grasped by information technology, is game-based based e-learning (Handri et al., 2010).

With the enrichment in technology, gamification with e-learning is a fruitful technique to engage the learners. Gamification is not about only playing games only but applying thinking of game-design into non-gaming applications (Ding et al., 2019). The use of information technology in the field of education plays an imperative role in learning technology development. Instead of education organizations, e-learning also gets attention in companies to train their workers. However, the main focus of e-learning is only on content delivery rather than how the user interacts (Troussas et al., 2020). Though this technique is working well but it should also cover some more prospective like how the learner interacts and how much the content is appealing to learner rather than onlyfocusing on content delivery (Wang et al., 2018). Mobile game-based learning shows the positive results in students' learning execution. With the new innovative turns of events, new strategies for instructing through mobiles are emerging. The utilization of versatile games in learning is a productive mix of dynamic and suited learning with fun in a conceivably amazing way (Prasolenko et al., 2019).

*Figure 1. Game-based learning*

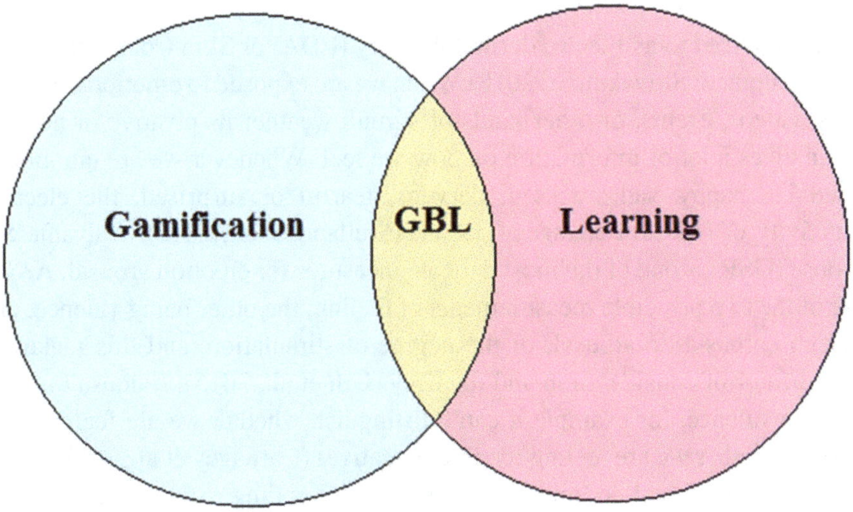


## What are the Benefits of E-Learning with Gamification?

1. Better Learning Experience
   The learner experiences more fun with more engagement in games. The can retain more information with high engagement in game (Jagušt et al., 2018).
2. Better Learning Environment
   Gamification with e-learning provides a space to the user where he/she can practice real life challenges in an informal and safe environment. More engagement in game leads to more retention of information (Sharma et al., 2018).
3. More Satisfaction and Motivation
   Student's satisfaction, motivation, engagement and efficiency can be increased by model gamification with e-learning. Effective learning suggests that the learning is effective if it is experimental, active; problem based and provides instant feedback (Naveen Kumar & Shivakumar, 2018).
4. Prompting Behaviour Change
   Point, badges and score sharing with friends makes the learning more joyful. However, gamification is a lot more than surface benefits (Udovičić et al., 2017). When combined with the scientific principles of repeated retrieval and spaced repetition, gamification can provide strong behavioral changes (Sharma et al., 2017).

## GALVANIC SKIN RESPONSE

GSR also referred to as Electro dermal Activity (EDA) or Skin Conductance (SC) (Naveen Kumar & Shivakumar, 2018).When we are exposed to emotionally loaded images, videos, events, or other kinds of stimuli weather its positive or negative, our skin gives a lot of information on how we feel. Whenever we are emotionally aroused i.e. happy, sad, stressed, nervous, fearful or surprised, the electrical conductivity of our skin changes (Gábana Arellano et al., 2016). Galvanic Skin Response (GSR) is one of the most delicate measures for emotion arousal. Arousal is one of the two principle measurements of feeling, the other being valence. GSR can catch enthusiastic arousal, or the degree of stimulation, and this makes it a solid indicator of consideration and memory (Urh et al., 2015). It doesn't uncover passionate valence, for example it can't distinguish whether we are feeling better (good) or whether we are feeling terrible (negative) (Carnagey et al., 2007). In this regard, GSR and facial coding are integral. Facial coding catches the valence of feeling, however can't measure the related arousal (imotions, 2021).

To know how GSR works, we need to have a look on the physiological characteristics of the skin, which is the largest organ of human body (Neulog, 2021).

- Skin acts as a protective barrier which separates our body from the environment and protects our body organs from the environments impacts such as temperature variations, radiations, chemical agents, pressure etcetera (Tobii Pro,).
- Our body temperature is controlled by the skin by regulating emission of sweat, goose bumps and peripheral blood circulation (Maulsby & Edelberg, 1960).
- Skin is an extensive network of nerve cells which can detect any change in the environment on the activity of the receptors for temperature, pain and pressure (ElProCus, 2021).

Our body contains millions of sweat glands whose density varies across the body. Sweat glands are being highest on cheeks and forehead, fingers and palms and as well as on the sole of feet (imotions, 2021). Whenever the sweat glands are triggered and becomes active, they secret moisture through the skin pores (Sonali, 2020). Variation in the balance of the positive and negative ions of the secreted fluid results in the flow of electrical current more readily which results in the measurable changes in the skin conductance (Brain Signs,). When the skin conductance increases, the skin resistance decreases. The emotional experience triggers the change in the arousal level which ultimately increases the sweat secretion and our heart beat increases and our hands become sweaty (Villarejo et al., 2012).

*Figure 2. Basic emotions*

*Figure 3. GSR sensor*

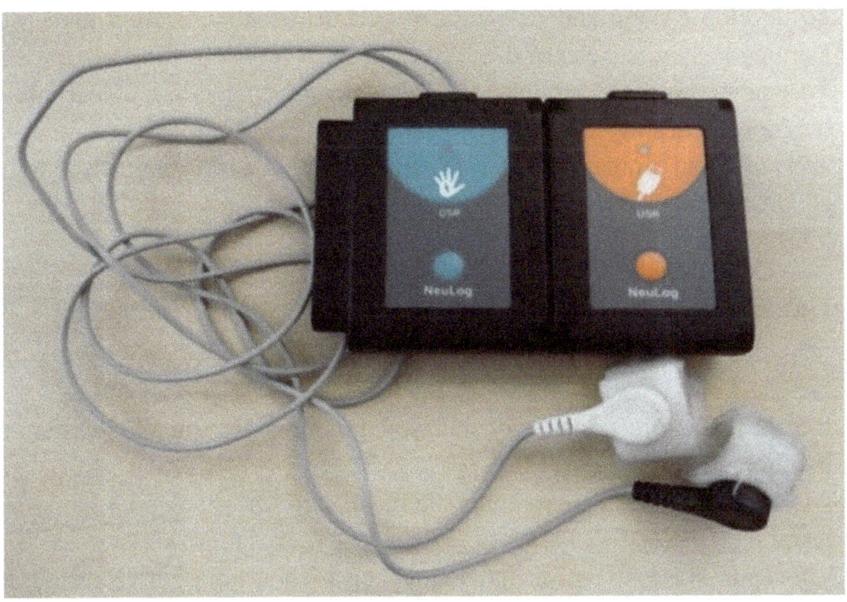

## FINDINGS OF LITERATURE SURVEY

After investigation of previous investigations, following problems are evaluated which are briefly discussed below.

1. Gamification with e-learning has a major focus on content delivery rather than learner's interaction (de-Marcos et al., 2016).
2. Gamification with e-learning uses short intervention time in single session (Barlow & Hayes, 1979). The experiments are conducted on the all the participants at same time within short time interval of 8-10 minutes (Abdul Jabbar & Felicia, 2015).
3. Efficacy of game-based learningis notdetermined in other subjects areas as well (e.g. language learning) (Hwang & Chang, 2015). Since the experiments are done on only mathematics functions i.e. addition and subtraction, vocabulary or language section can also be included to enhance the findings (Tijs et al., 2008).

## END GOAL

Following are the various intentions of this research work.

1. To develop an app and design an experiment for inculcating game-based learning in a typical classroom.
2. To conduct various sessions for implementing game-based learning and traditional learning.
3. To analyse and compare the effect of game-based learning in typically developed children through physiological signals.

## ASSUMPTIONS

1. It had been assumed that students' scored more in game-based learning than traditional learning.
2. Also, students are likely to complete the task in less time under game-based learning.
3. The captured physiological signals are likely to show higher arousal in students of game-based learning rather than traditional learning.

## STRATEGY

1. In this research work, parameters to develop a game will be selected.
2. After game development, experiment has to be conducted for implementing game-based learning as well as traditional learning. There have been various levels whose difficulty will be increased with each level increment. Number of students are 40 of age group 6-8 years. The students have been divided into two groups i.e. Group A and Group B. Group A students undergo game-based learning and Group B students under traditional learning.
3. Different physiological signals have been acquired while performing experiment. Data was collected during experiment which will be processed and analyzed using MATLAB software. Since the collected data will contain a lot of noisy components due to vibrations and fluctuations in any electronic component while collecting the data, it is important to filter our signal to suppress artifacts. For this purpose, moving average filter has been applied to suppress noise.

*Figure 4. Strategy flow chart*

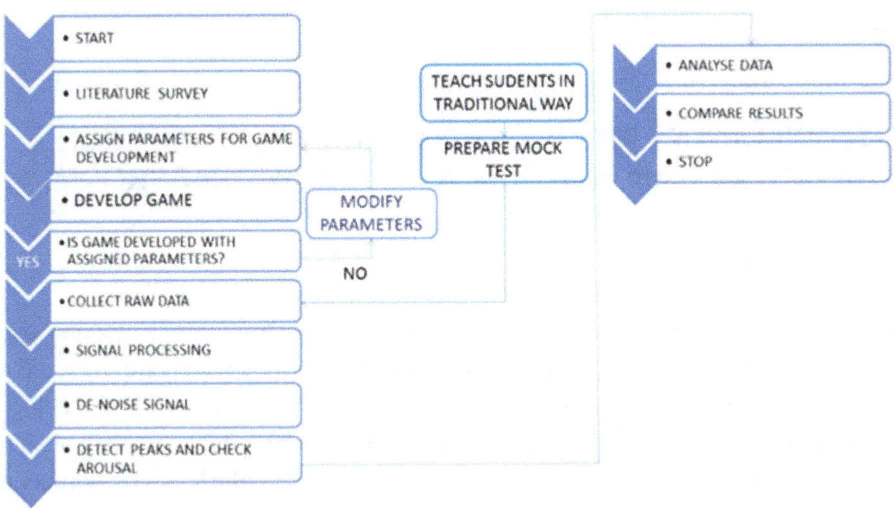


## DATA COLLECTION

The physiological signals have been recorded for each student in both game play learning and traditional learning using GSR for sampling rate 20 samples per second. Firebase console has been used to store the information about each level for student playing game. The data was stored for each level giving information about the correct and incorrect answers, time and total scores for each different levels of the student under English and Math level. The information of each student has been displayed under the assigned ID. The data collected in this research has leveraged using two data sources for both game-based learning and traditional learning (a) student-generated data (b) pre and post-experiment interviews. Data for students in traditional learning will be recorded manually in which the time will be recorded for each student along with correct- incorrect attempts and total scores.

*Figure 5. Database console*

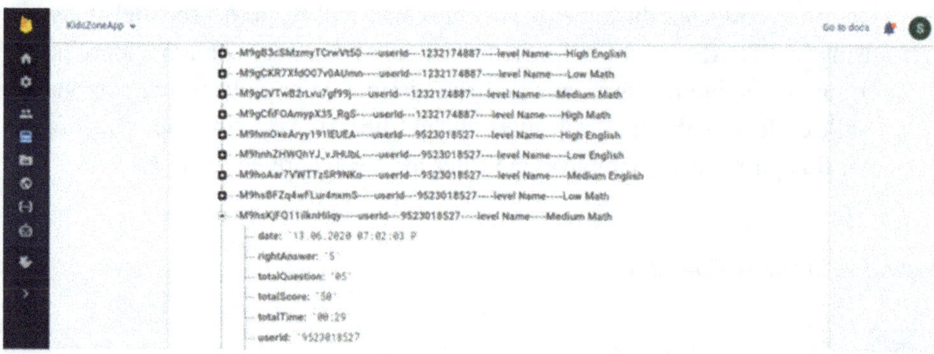

## TO DEVELOP AN APP AND DESIGN AN EXPERIMENT FOR INCULCATING GAMIFICATION IN A TYPICAL CLASSROOM.

### Game Design

To conduct the experiment, a custom mobile learning platform called "KidsZoneApp" has been developed which is simple yet media-rich to provide the better interaction to the students. Different Ids has been generated which will be given to each student playing game. This application is used to perform experiment on Group A students. The game design included two levels – Math and English. Each of the levels comprehends three levels further i.e. low, medium and high. This game platform has been used to solve three mathematics operations i.e. Addition, Subtraction and Multiplication. Furthermore, English section contains scrambled words and vocabulary section. For every correct answer, 10 points are rewarded. For every correct answer given by the student, happy face emoji appeared on the screen. Correspondingly, sad face emoji appeared for every wrong answer. However, there is no negative marking for wrong answers. For the purpose of the experiment, each module is of 9 minutes;

which means the total game-play will last for 18 minutes. During the game play learning, level, total scores, time left and question number has been mentioned at left and right corner of the device screen. At the end of every level, score board appears giving information about total correct, total incorrect, total time and total score for each level of student performance. Before the experimental procedure, the students will be taught how to use the game and touch the screen.

**Experimental Results:** Figure 6 shows the screen captures of the home screen of KidsZoneApp. At the point when we enter the given ID, this home screen opens where we can choose the module, for example, English or Math. After the choice of any module, various levels open up for example low, medium, and high. By the determination of any level, the questions allocated to specific level come up.

*Figure 6. Home screen of KidsZoneApp*

*Figure 7. Screenshots of math module from KidsZoneApp*

Figure 7 shows the screen captures of the Math module while figure 8 shows the screen captures from the English module for all the various levels. In the two cases, we can see the level number at the head of the screen. The upper left corner of the screen shows the scores acquired by an understudy alongside the question number an understudy is endeavoring. The upper right corner of the screen gives data about the time left to finish the level.

*Figure 8. Screenshots of english module from KidsZoneApp*

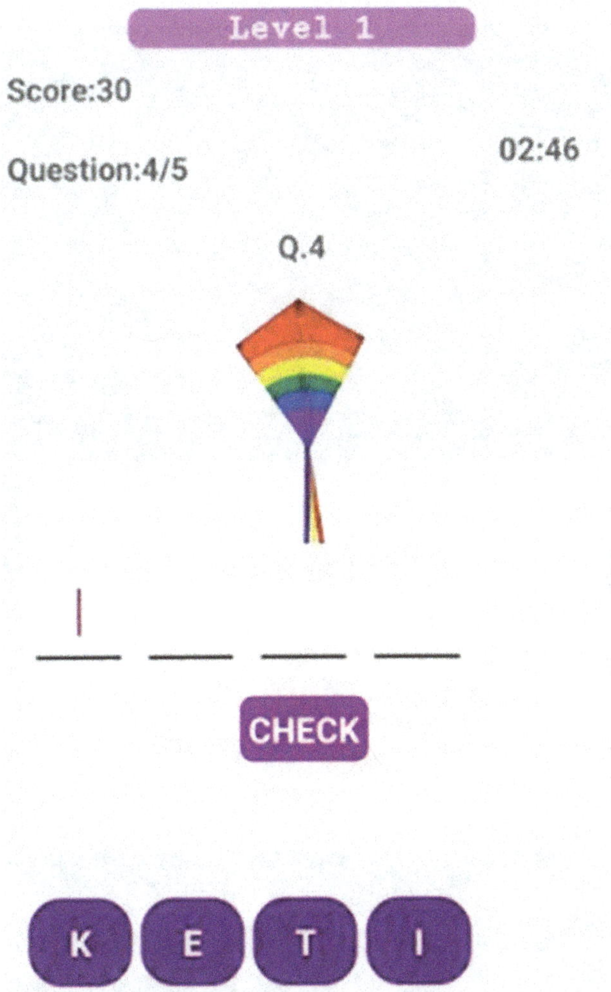


Figure 9 shows the pop-ups that showed up on screen after every question attempt. For every correct answer, cheerful face pops-up and for each wrong answer, miserable face pops-up.

Figure 10 shows the screen capture of the result shown on screen after every level attempt. It gives the information about number of question composed in level, correct attempts, incorrect attempts, time taken and total scores obtained in the level.

*Figure 9. Screenshots of happy and sad faces for correct and incorrect answers respectively*

*Figure 10. Screenshot of result shown after each played level*

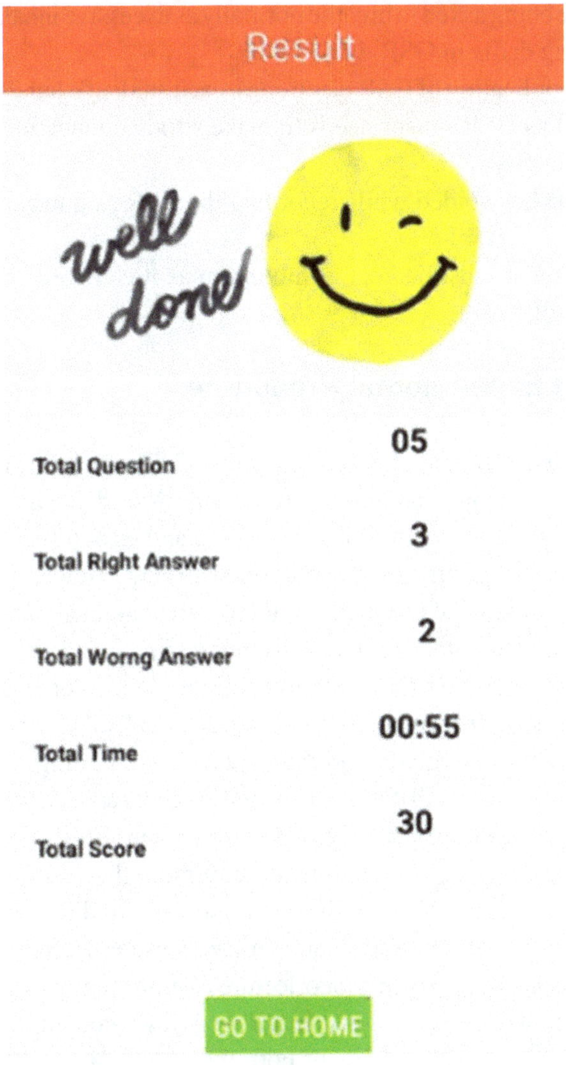


## TO CONDUCT VARIOUS SESSIONS FOR IMPLEMENTING GAMIFICATION

### Considerations for Conducting Experiment

Before starting with the experiment, students of both the groups have given certain guidelines in order to perform the experiment. These guidelines are as follow.

1. Student must sit comfortably during the experiment.
2. Students are being asked to breathe normally as excessive inhalation/exhalation causes strong drifts in GSR data.
3. Before the application of GSR electrodes, each student's hands are sanitized by sanitizer gel as GSR sensor needs to make strong connection to skin in order to works properly.
4. Talking must be avoided while recording the GSR as it may cause artifacts in signal.
5. Limb movement must be kept at minimum as it can comprehensively affect the GSR data.

## Implementing Experimental Procedure

Number of students participated in this study are 40. The students has been divided into two groups i.e. Group A and Group B. Groups A students has been those who have has a prior exposure of game-based learning and Group B students have been those who have no exposure of game-based learning. Before starting with the procedure, all the students of group A and B has been taught about the mathematics operations and English sentences. The session lasted for 2-3 days for every student. Parents were asked to give training to their children. For this process, an inbuilt app named "KidsZoneApp" from Google Play Store is downloaded for Group A students and Group B students were given traditional paper-pen training.

Starting with experiment, all the experimental procedures occurred in a soundproof room. The room temperature has been kept between 20-40°C; since the GSR information can be dissipated at a temperature of more than 40°C and under 20°C. The participants sat comfortably on the chair and were told to stay still in order to not move the wires of the electrodes since it can cause artifacts to the signal. The group A students has been given video lectures before the game play and on the contrary, the Group B has been taught in traditional way before the experiment. Time limit has been known to the students of both the groups A and B. Before starting the experiment, the pre-experiment baseline has been recorded for each student for the duration of 10 seconds. After that, signals were acquired during experimental protocol at sampling rate 20 samples/ second. Post-experiment arousal of each student has been recorded for 10 seconds for both the groups A and B. The time interval between baseline signal acquisition before and after the experiment was 5 minutes. In addition to this, pre-experimental and post-experimental interview has been conducted respectively before and after data collection. The pre-experimental interview lasted for 5 minutes in which each of the participants was asked general questions about their name, age, class, preferred hand and weather they prefer game play learning or traditional learning. Again, post-experimental interview has been lasted for 5 minutes in which participant was asked whether they were stressed, did they guessed any answer and did they enjoyed game.

## Experimental Results

*Figure 11. Students having game-based learning*

Figure 11 shows the students of Group A, who undergoes game-based learning. While carrying this experimental procedure, GSR sensor electrodes have been attached to the non-preferred hand of students. Also, electrodes wires must not dangled loosely in order to avoid artifacts. The room temperature has been also monitored continuously in order to acquire the data between described temperatures. Each student has been acquired by one of their parents in order to increase the comfortable level.

*Figure 12. Students having traditional learning*

Figure 12 shows the students of Group B who undergoes traditional learning. While carrying this experimental procedure, GSR sensor electrodes were attached to the non-preferred hand of students. Also, electrodes wires were not dangled loosely in order to avoid artifacts. The room temperature was also monitored continuously in order to acquire the data between described temperatures. Each student was acquired by one of their parents in order to increase the comfortable level.

*Figure 13. Screen capture of temperature monitoring during experiment*

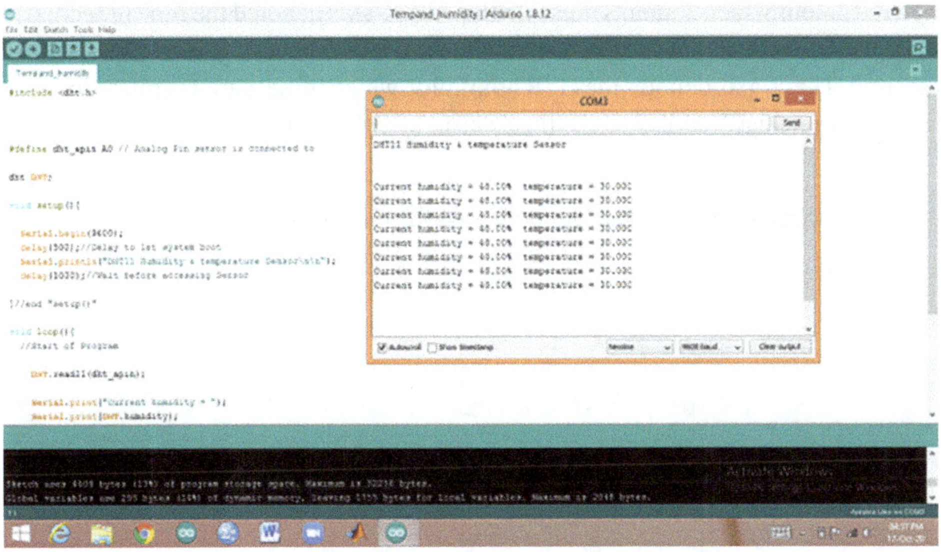


## TO ANALYZE AND COMPARE THE EFFECT OF GAMIFICATION IN TYPICALLY DEVELOPED CHILDREN THROUGH PHYSIOLOGICAL SIGNALS

The major dependent variable is SCR, which has been recorded at real-time during experimental protocol.Since, the real time SCR includes a lot of noise due to power hindrance or vibrations from electronic devices like air conditioner or lights etc (Salomon, 1984). Therefore, is important to de-noise the acquired physiological signal in order to obtain useful data.For this purpose, MATLAB software has been used in which the GSR data was loaded and then moving average low pass filter has been applied to each acquired signal. After that, maximum peak for each signals were detected and then, average of all maximum peaks of were calculated for each participant both groups A and B.

After that, ANOVA (analysis of variance) has been applied to both the groups in order to check if there is any significant difference in both the groups A and B. The ANOVA has been applied for the overall comparison of both groups A and B and as well as subject wise comparison between groups A and B. The comparison is made on the basis of points scored and time taken by the students of the respective group. At last, ANOVA is applied to the mean peak arousal of both the groups.

## Experimental Results

Figure 14 shows the screen capture of the raw baseline signal of a student represented by a red colour which contains noise. In this, the X-axis denoted the time in seconds; since the baseline signal is captured for 10 seconds at a sampling rate of 20 samples / second. The Y-axis denotes the GSR amplitude which is measured in microsimens; or we can say that Y-axis denotes the arousal of GSR data.

*Figure 14. Raw baseline signal uploaded on Matlab software*

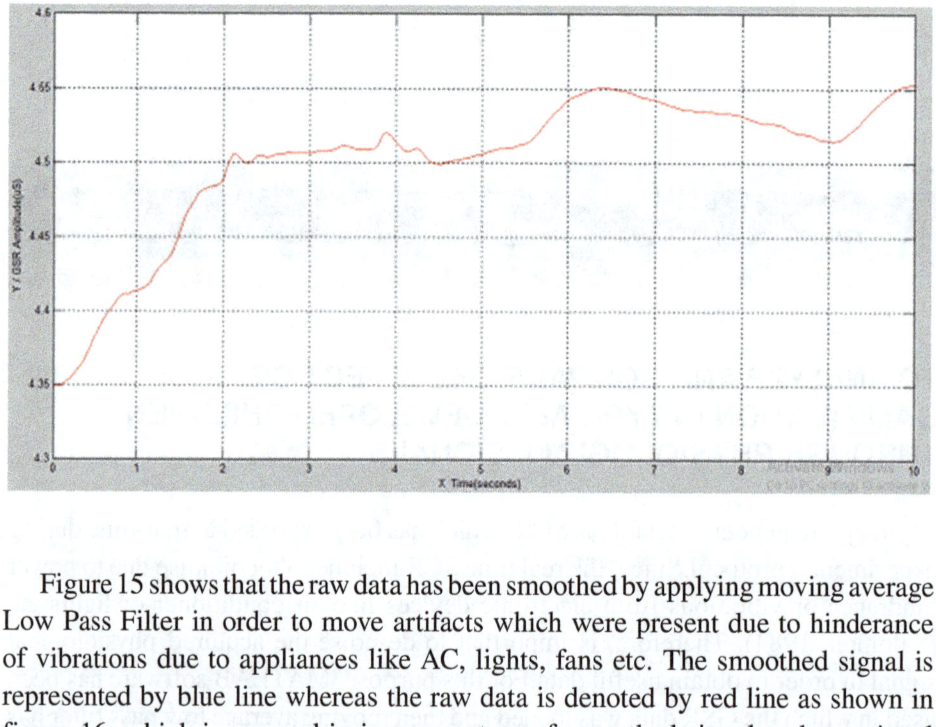


Figure 15 shows that the raw data has been smoothed by applying moving average Low Pass Filter in order to move artifacts which were present due to hinderance of vibrations due to appliances like AC, lights, fans etc. The smoothed signal is represented by blue line whereas the raw data is denoted by red line as shown in figure 16; which is zoomed in image capture of smoothed signal and raw signal captured by GSR sensor.

*Figure 15. Smoothed signal*

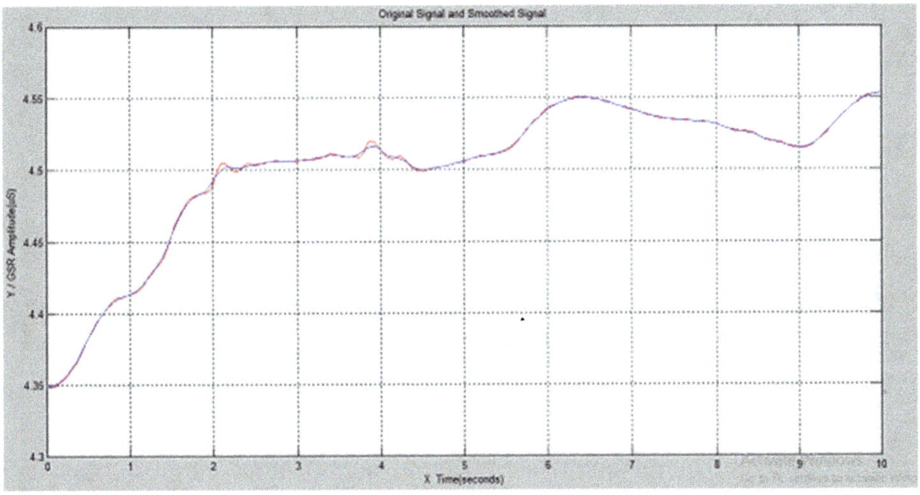


*Figure 16. Zoomed in capture of smoothed signal denoted in blue colour*

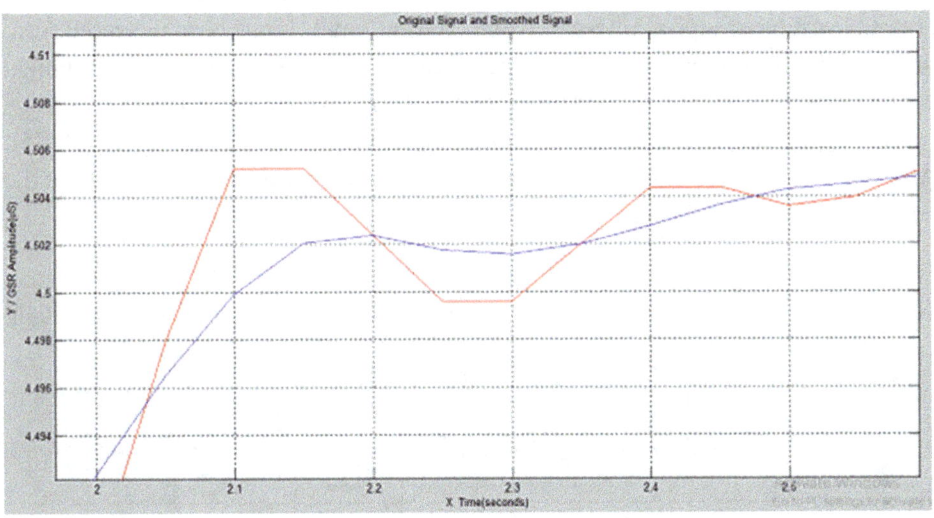


Figure 17 shows the screen capture of the processed signal in which red signal depicts the original signal which contains noise, whereas blue signal is a de-noised signal. After de-noising, the peaks is identified which is pointed by asterisk (*) in green colour.

*Figure 17. Peak detection from smoothed signal*

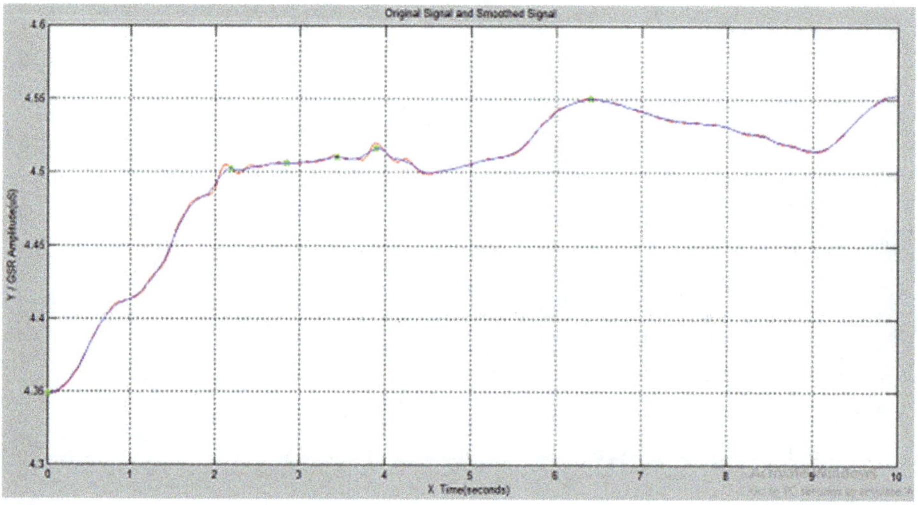

## RESULTS

**Experimental Results:** Figure 18 shows the average score and time as a function of both learning groups (game-based and conventional). It has been observed from the Table 1 that both the groups show significant difference in mean marks obtained by the students. ANOVA yielded an exceptionally noteworthy distinction in average scores of game-based learning and traditional learning ($F(1, 38) = 8.168$, $p<0.05$). Also, ANOVA produce a critical distinction between the time produced for both the learning ($F(1, 38) = 36.525$, $p< 0.001$).

*Table 1. ANOVA results of comparison of GBL and TL as a function of scores and time*

| Descriptive | | | | | | | | | |
|---|---|---|---|---|---|---|---|---|---|
| | | N | Mean | Std. Deviation | Std. Error | 95% Confidence Interval for Mean | | Minimum | Maximum |
| | | | | | | Lower Bound | Upper Bound | | |
| Scores | Gamified Learning | 20 | 236.5000 | 25.60325 | 5.72506 | 224.5173 | 248.4827 | 200.00 | 280.00 |
| | Traditional Learning | 20 | 202.0000 | 47.52838 | 10.62767 | 179.7560 | 224.2440 | 120.00 | 270.00 |
| | Total | 40 | 219.2500 | 41.53389 | 6.56708 | 205.9668 | 232.5332 | 120.00 | 280.00 |
| Time | Gamified Learning | 20 | 263.1500 | 48.27910 | 10.79553 | 240.5547 | 285.7453 | 160.00 | 329.00 |
| | Traditional Learning | 20 | 564.5500 | 217.74187 | 48.68856 | 462.6437 | 666.4563 | 225.00 | 948.00 |
| | Total | 40 | 413.8500 | 218.00524 | 34.46965 | 344.1285 | 483.5715 | 160.00 | 948.00 |

*Table 2.*

| ANOVA | | | | | | |
|---|---|---|---|---|---|---|
| | | Sum of Squares | df | Mean Square | F | Sig. |
| Scores | Between Groups | 11902.500 | 1 | 11902.500 | 8.168 | .007 |
| | Within Groups | 55375.000 | 38 | 1457.237 | | |
| | Total | 67277.500 | 39 | | | |
| Time | Between Groups | 908419.600 | 1 | 908419.600 | 36.525 | .000 |
| | Within Groups | 945105.500 | 38 | 24871.197 | | |
| | Total | 1853525.100 | 39 | | | |

*Figure 18. Overall comparison*

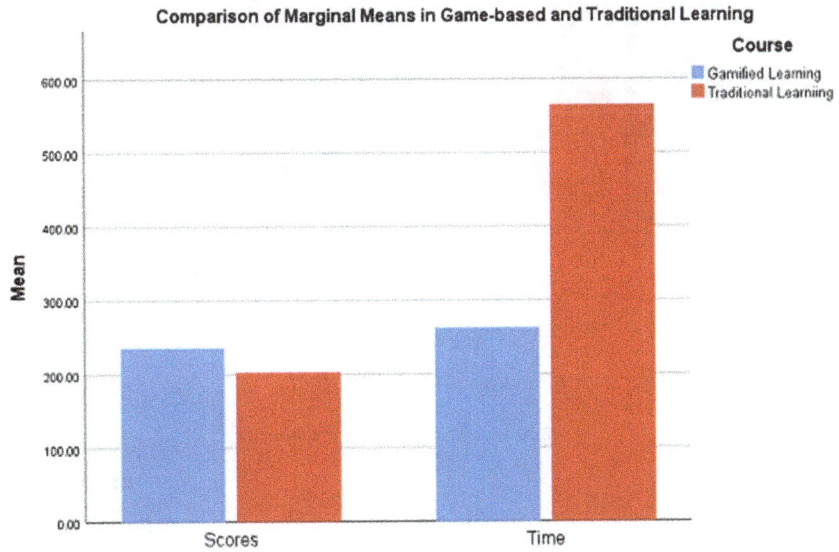


By examining the mean scores and time of both the subjects as an element of groups (game-based and learning), figure 19 and figure 20 shows the average scores and time obtained respectively in both the subject; English and Math; as a function of two learning (game-based and traditional).The ANOVA yielded a significant difference in both groups for variables average time and scores. The mean score of English module showed significant difference (F (1.38) = 1.780, p <=1.0) and for Math (F= (1, 38) =11.753, p=0.001). Similarly for time, ANOVA showed significant difference (F (1, 38) = 32.563, $p < 0.001$) for English as well as for Math (F (1, 38) = 28.417, $p < .001$).

Thus we can say that, on the basis of ANOVA results, students with game-based learning scored more in lesser time as compared to traditional learning.

*Figure 19. Mean scores obtained for both the subjects*

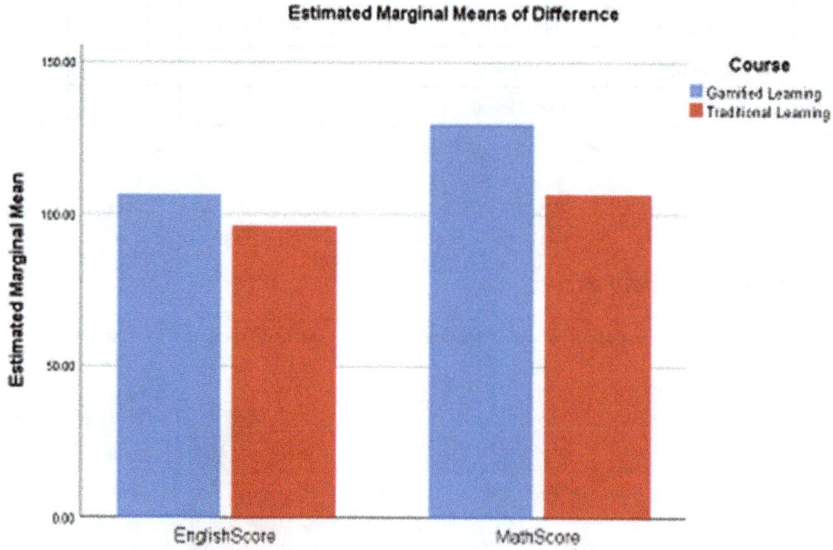


*Figure 20. Mean scores obtained for both the subject*

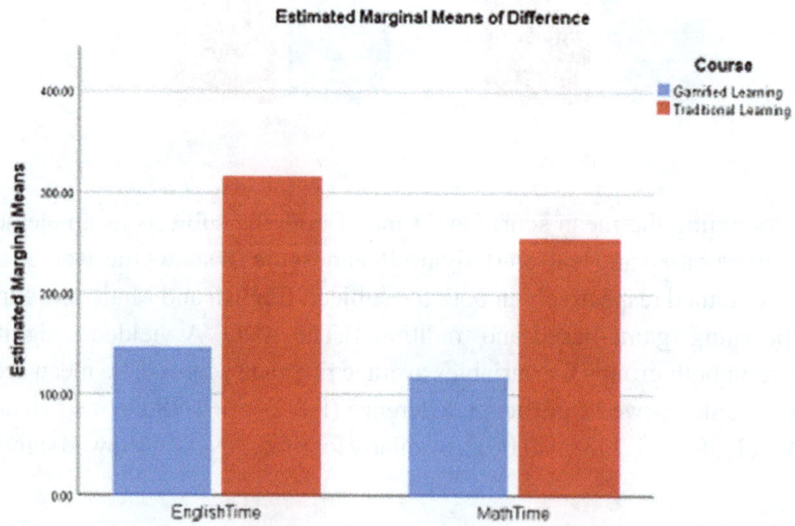

*Table 3. ANOVA results of subject-wise comparison of GBL and TL as a function of Scores obtained*

| Descriptive | | | | | | | | | |
|---|---|---|---|---|---|---|---|---|---|
| | | N | Mean | Std. Deviation | Std. Error | 95% Confidence Interval for Mean | | Minimum | Maximum |
| | | | | | | Lower Bound | Upper Bound | | |
| English Score | Gamified Learning | 20 | 106.5000 | 17.85173 | 3.99177 | 98.1451 | 114.8549 | 80.00 | 130.00 |
| | Traditional Learning | 20 | 96.0000 | 30.33150 | 6.78233 | 81.8044 | 110.1956 | 50.00 | 150.00 |
| | Total | 40 | 101.2500 | 25.13425 | 3.97407 | 93.2117 | 109.2883 | 50.00 | 150.00 |
| Math Score | Gamified Learning | 20 | 130.0000 | 12.13954 | 2.71448 | 124.3185 | 135.6815 | 100.00 | 150.00 |
| | Traditional Learning | 20 | 106.5000 | 28.14904 | 6.29432 | 93.3258 | 119.6742 | 50.00 | 150.00 |
| | Total | 40 | 118.2500 | 24.48312 | 3.87112 | 110.4199 | 126.0801 | 50.00 | 150.00 |

*Table 4.*

| ANOVA | | | | | | |
|---|---|---|---|---|---|---|
| | | Sum of Squares | df | Mean Square | F | Sig. |
| English Score | Between Groups | 1102.500 | 1 | 1102.500 | 1.780 | .100 |
| | Within Groups | 23535.000 | 38 | 619.342 | | |
| | Total | 24637.500 | 39 | | | |
| Math Score | Between Groups | 5522.500 | 1 | 5522.500 | 11.753 | .001 |
| | Within Groups | 17855.000 | 38 | 469.868 | | |
| | Total | 23377.500 | 39 | | | |

From the results obtained, we can say that both learning have a significant difference in which game-based learning outshined traditional learning. Also, there is a major difference between the average time taken by both the groups in the completion of task. Besides, in subject-wise examination also, game-based learning surpassed conventional learning as an element of score and time.

Figure 21 shows the SCR response for pre and post-experiment baseline for 12 students randomly choose from both the groups. 6 students (A-F) have been randomly selected from Group A and 6 students (G-L) have been randomly selected from Group B. It has been clearly seen that the arousal of all the participants increased after an experimental protocol. ANOVA yielded a very significant response in pre and post- experiment baseline (F (1, 22) = 3.29043, p<0.1).

*Table 5. ANOVA results of subject-wise comparison of GBL and TL as a function of time taken*

| | | N | Mean | Std. Deviation | Std. Error | 95% Confidence Interval for Mean | | Minimum | Maximum |
|---|---|---|---|---|---|---|---|---|---|
| | | | | | | Lower Bound | Upper Bound | | |
| English Time | Gamified Learning | 20 | 147.1000 | 43.27258 | 9.67604 | 126.8478 | 167.3522 | 60.00 | 240.00 |
| | Traditional Learning | 20 | 315.7500 | 124.88684 | 27.92555 | 257.3012 | 374.1988 | 127.00 | 552.00 |
| | Total | 40 | 231.4250 | 125.71271 | 19.87693 | 191.2201 | 271.6299 | 60.00 | 552.00 |
| Math Time | Gamified Learning | 20 | 118.1500 | 28.61040 | 6.39748 | 104.7599 | 131.5401 | 55.00 | 173.00 |
| | Traditional Learning | 20 | 254.6500 | 110.88271 | 24.79413 | 202.7553 | 306.5447 | 85.00 | 497.00 |
| | Total | 40 | 186.4000 | 105.66997 | 16.70789 | 152.6051 | 220.1949 | 55.00 | 497.00 |

*Table 6.*

| ANOVA | | | | | | |
|---|---|---|---|---|---|---|
| | | Sum of Squares | Df | Mean Square | F | Sig. |
| English Time | Between Groups | 284428.225 | 1 | 284428.225 | 32.563 | .000 |
| | Within Groups | 331915.550 | 38 | 8734.620 | | |
| | Total | 616343.775 | 39 | | | |
| Math Time | Between Groups | 186322.500 | 1 | 186322.500 | 28.417 | .000 |
| | Within Groups | 249157.100 | 38 | 6556.766 | | |
| | Total | 435479.600 | 39 | | | |

*Figure 21. SCR Response for Pre and post-experimental baseline*

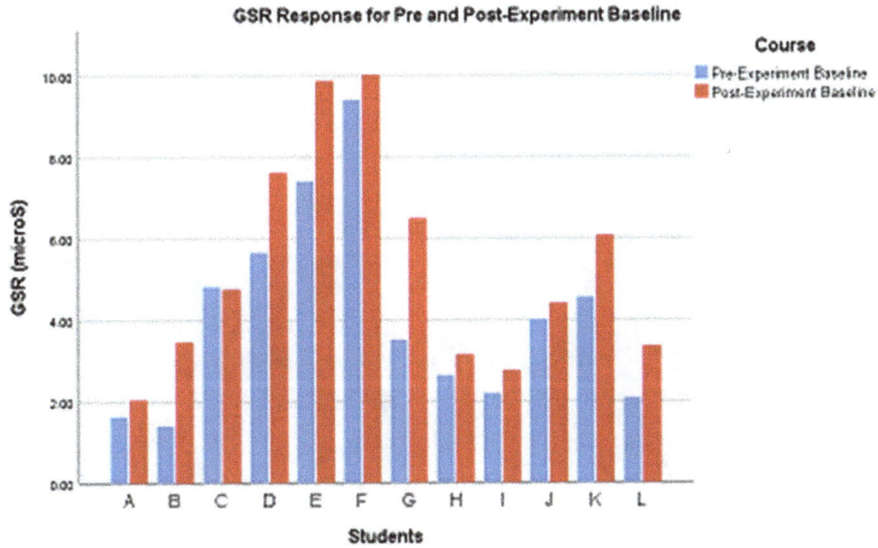


The figure 22 below shows the average SCR response of students in both the learning (game-based and conventional). ANOVA yielded a very significant response in both the groups (F (1, 38) =11.424, p<0.1) showing that there is an enormous distinction in SCR reaction of both the gatherings A and B down the test system.

*Table 7. ANOVA results of comparison of SCR in GBL and TL*

| **Descriptive** | | | | | | | | |
|---|---|---|---|---|---|---|---|---|
| **Arousal** | | | | | | | | |
| | **N** | **Mean** | **Std. Deviation** | **Std. Error** | **95% Confidence Interval for Mean** | | **Minimum** | **Maximum** |
| | | | | | **Lower Bound** | **Upper Bound** | | |
| Game-based Learning Arousal | 20 | 27.4434 | 12.55561 | 2.80752 | 21.5672 | 33.3196 | 5.87 | 41.39 |
| Traditional Learning Arousal | 20 | 17.1483 | 5.28357 | 1.18144 | 14.6755 | 19.6211 | 9.30 | 25.01 |
| Total | 40 | 22.2959 | 10.84331 | 1.71448 | 18.8280 | 25.7637 | 5.87 | 41.39 |

| **ANOVA** | | | | | |
|---|---|---|---|---|---|
| Arousal | | | | | |
| | Sum of Squares | df | Mean Square | F | Sig. |
| Between Groups | 1059.891 | 1 | 1059.891 | 11.424 | .002 |
| Within Groups | 3525.629 | 38 | 92.780 | | |
| Total | 4585.520 | 39 | | | |

*Figure 22. SCR response of students of group A and B*

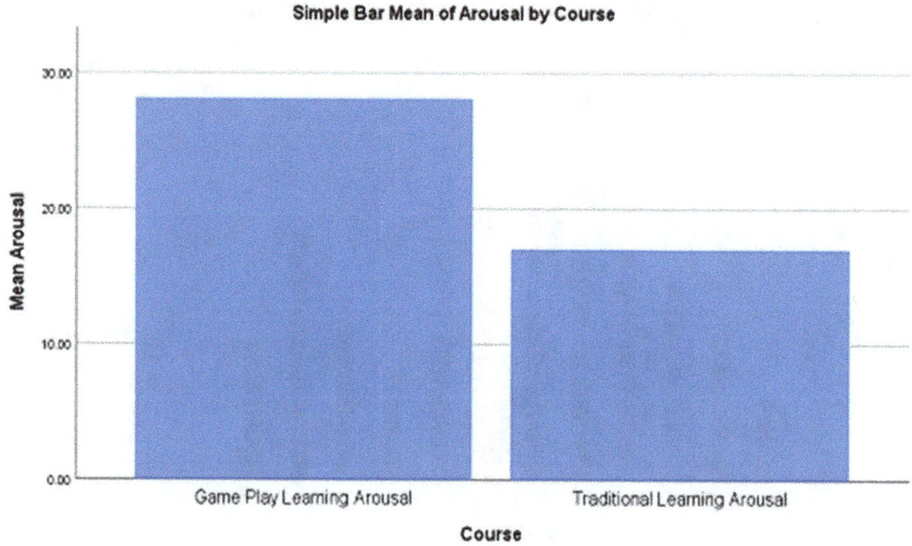


Figure 23 (a), (c), and (e) portrays the pre-experiment baseline signals of three different students, and figure 4.6 (b), (d), and (f) depicts the post-experiment baseline signals of the same three students respectively. It tends to be seen that the SCR increments after game intercessions. In figure 23 (a), (c), and (e); the mean peak arousal is 2.08, 4.55, and 9.39 in pre baseline whereas the arousal is 3.34, 6.04, and 9.99 in post-experiment baseline as shown in figure 23 (b), (d) and (f). This is because of the well-established fact that the physiological arousal continually changes for the duration of the day, and this variance significantly impacts conduct and execution (Nozawa & Uchida, 2009). Variances in excitement are regularly connected to changes in the sympathetic and parasympathetic activity in the autonomic nervous system. Galvanic skin reaction (GSR) is free of sympathetic action, while the parasympathetic sensory system prevalently constrains the heart rate (HR) (Scotti, 2006). The sympathetic nervous system controls sweat gland activity, and expansion in a sympathetic activity produces a corresponding increase in GSR (Wang et al., 2004).

*Figure 23. (a), (c) and (e) pre experiment baseline signals and (b), (d) and (f) post experiment baseline signals*

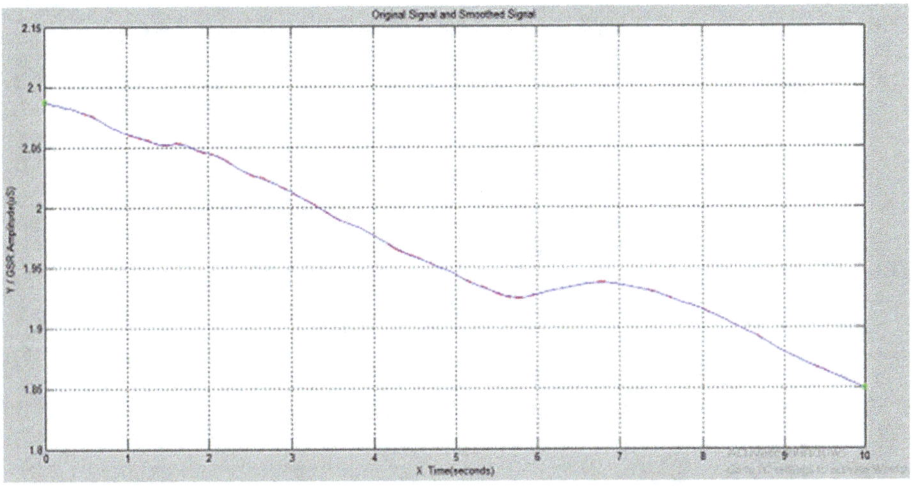


Figure 24 are the signals of two students during experiment. Figure 24 (a) and (c) are the signals obtained for English level whereas figure 24 (b) and (d) are signals obtained for Math level. These are the signals of the students who scored well in the experimental protocol. The marks obtained by both the students in English module are 140 and 130 out of 150 and of Math module are 110 and 140 out of 150 respectively. The SCR response of first student is 5.38 for English module and 6.18 for Math module whereas the SCR response of second student is 7.23 for English module and 7.40 for Math module.

*Figure 24. Signals of two students from group A where (a), (c) are of English level and (b) and (d) are of Maths level*

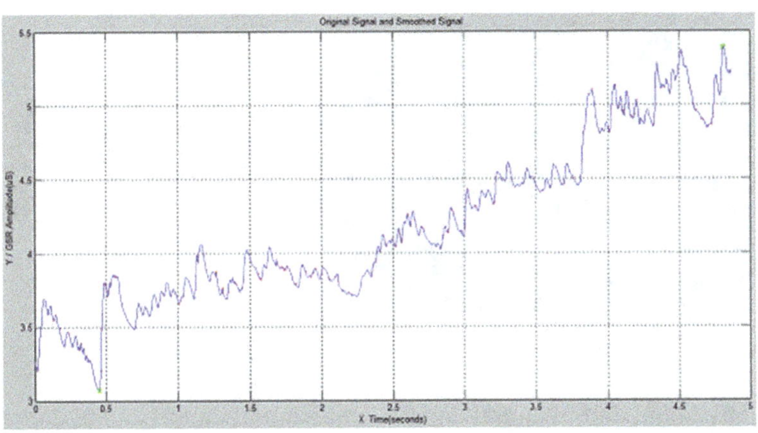

*Figure 25. Signals of two students from Group B where (a), (c) are of English level and (b) and (d) are of Maths Level*

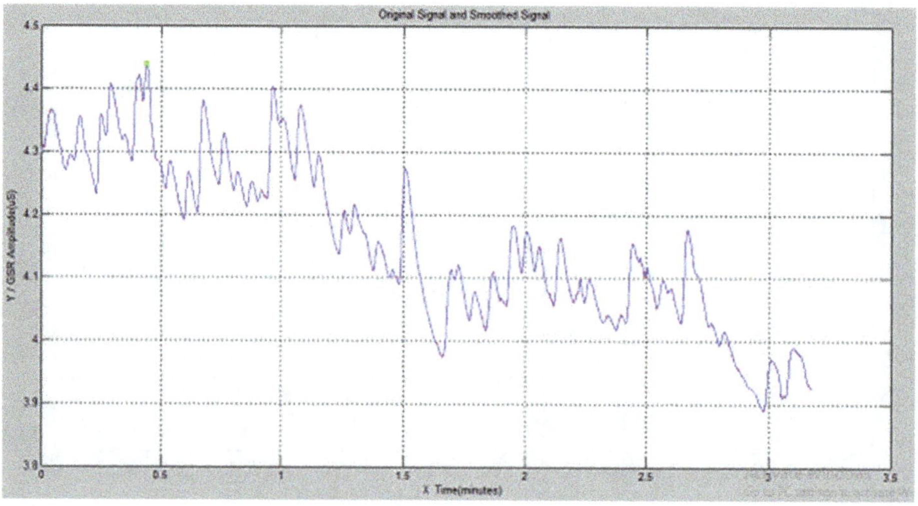

Figure 25 are the signals of two students of Group A captured during experiment. Figure 25 (a) and (c) are the signals obtained for English level whereas figure 25 (b) and (d) are signals obtained for Math level. These are the signals of the students who scored less in the experimental protocol. The marks obtained by both the students in English module are 90 and 70 out of 150 and of Math module are 120 and 120 out of 150 respectively. The SCR response of first student is 4.43 for English module and 4.74 for Math module whereas the SCR response of second student is 4.12 for English module and 3.67 for Math module.

*Figure 26. Percentage of students enjoyed game-based learning (GBL)*

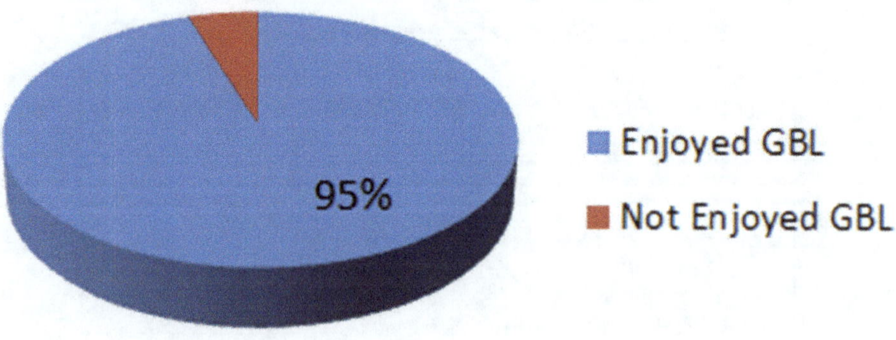

Figure 26 shows the percentage of students who really enjoyed the game-bassed learning (GBL) which is 95%. That means all the students of Group A enjoyed the GBL.

*Figure 27. Percentage of students who guessed or not guessed any answer game-based learning (GBL)*

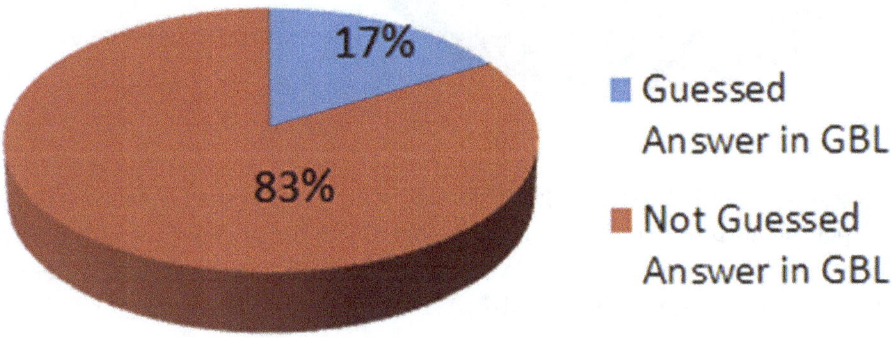


Figure 27 shows the percentage of students who guessed the answer during experimental procedure of GBL. It shows that 15% of the students guessed the answer while remaining 85% students didn't guessed any answer.

*Figure 28. Percentage of students who guessed or not guessed any answer traditional learning (TL)*

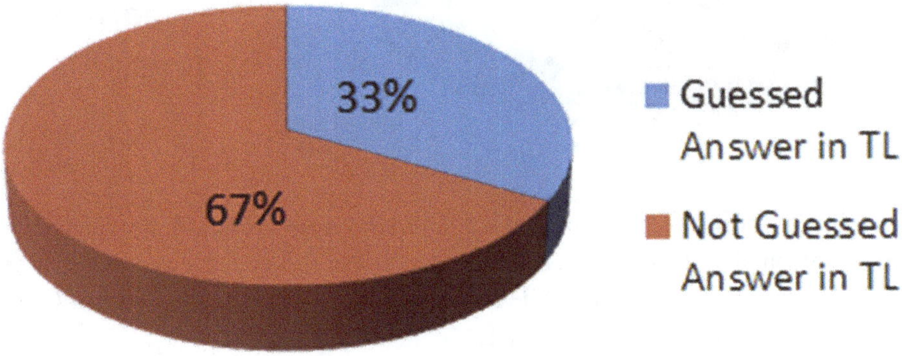


Figure 28 shows the percentage of students who guessed the answer during experimental procedure of TL. It shows that 33% of the students guessed the answer while remaining 67% students didn't guessed any answer in traditional learning.

*Figure 29. Percentage of students who were stressed or nervous during GBL*

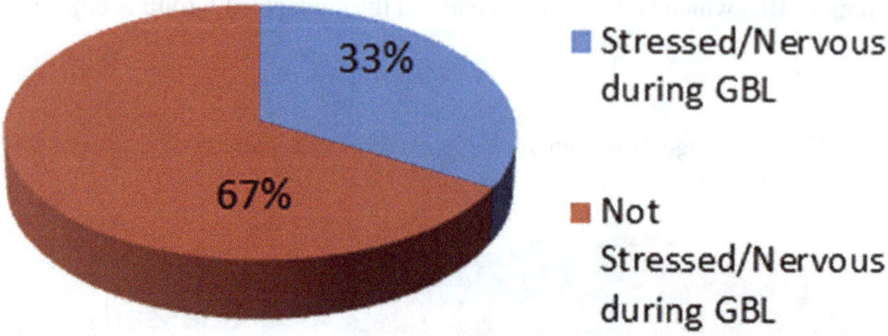


Figure 29 shows the percentage of students who were stressed or nervous before the experimental protocol of GBL. It shows that 67% of the students were not stressed or nervous during experiment, while remaining 33% of the total students of Group A were nervous or stressed under GBL.

*Figure 30. Percentage of students who were stressed or nervous during traditional learning (TL)*

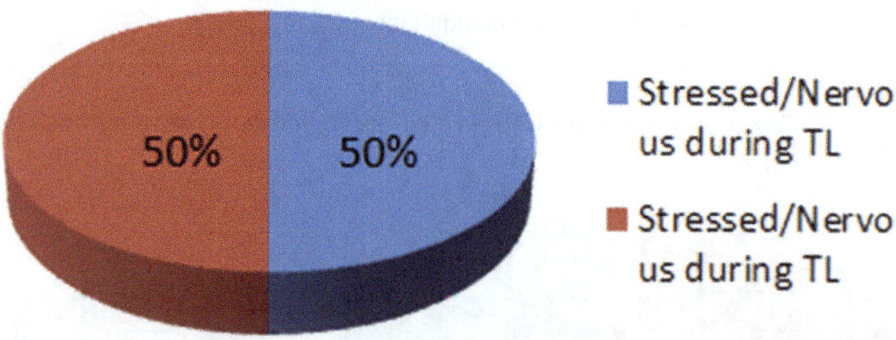


Figure 30 shows the percentage of students who were stressed or nervous before the experimental protocol of TL. It shows that 50% of the students were not stressed or nervous during experiment, while remaining 50% of the total students of Group A were nervous or stressed under GBL.

## CONCLUSION

The idea of this study was to inspect the change in the performance of game-based learning as compared to traditional learning. For this purpose, newly designed android based game intervention (KidsZoneApp) has been developed in order to provide finer engagement in student education. The children uncovered improvement under game-based learning. It had been observed that students under game-based learning gained immense scores and also they completed the task in minimum time. The children under Group A revealed higher average marks in both modules with marginal time. It had been observed in the results that Group A had higher SCR amplitudes in pre and post-baseline and in the experimental procedure as well. It had also been observed that using KidsZoneApp mobile-based platform; children are able to complete the task in a short time.

These results sustain the findings from previous studies that as a proportion of consideration, GSR identifies with learning. Various investigations discovered connections between learning and increased GSR. Understudies with higher GSR measures during learning scored best. Additionally, the knowledge level was identified with higher GSR measures, with more brilliant youngsters having a tendency to be increasingly responsive (higher GSR measures) (Werbach & Hunter, 2012).

A mobile-based game intervention with different subjects can enhance the learning-skills of students. Moreover, the game can help the students practice their skills out of the study hall. Students can practice sitting at homes or anywhere around the corner of the world and can practice new skills. The students have been additionally given vivified acclaim as 'smiley' for each right answer, which can inspire them to perform well.

## FUTURE SCOPE

1. The results of this work demands for the addition of more physiological signals such as ECG, blood pressure etc. in order to check the correlation between the SCR and other physiological signals.
2. The future work may also include exploring the concealed efficacy of incorrect attempts where students learn from their failed attempts (Yildirim, 2017).

## REFERENCES

Abdul Jabbar, A., & Felicia, P. (2015). Gameplay Engagement and Learning in Game-Based Learning. *Review of Educational Research*, *85*(4), 740–779. doi:10.3102/0034654315577210

Agrafioti, F., Hatzinakos, D., & Anderson, A. (2012). ECG Pattern Analysis for Emotion Detection. *IEEE Transactions on Affective Computing*, *3*(1), 102–115. doi:10.1109/T-AFFC.2011.28

Andreassi, J. L. (2007). *Psychophysiology: Human Behavior & Physiological Response* (5th ed.). Psychology Press.

Arroyo-Gómez, N., Laparra-Hernández, J., Soler-Valero, A., Medina, E., & de Rosario, H. (2017). *Physiological Model to Classify Physical and Cognitive Workload During Gaming Activities*. Advances in Neuroergonomics and Cognitive Engineering. doi:10.1007/978-3-319-60642-2_23

Ashok Charan. (n.d.). *Marketing analytics*. https://www.ashokcharan.com/Marketing-Analytics/~bm-galvanic-skin-response.php

Barlow, D., & Hayes, S. (1979). Alternating treatments design: One strategy for comparing the effects of two treatments in a single subject. *Journal of Applied Behavior Analysis*, *12*(2), 199–210. doi:10.1901/jaba.1979.12-199 PMID:489478

Brain Signs. (n.d.). *Sapienza University of Rome*. Retrieved from https://www.brainsigns.com/en/science/s2/tehnologies/gsr

Carnagey, N., Anderson, C., & Bushman, B. (2007). The effect of video game violence on physiological desensitization to real-life violence. *Journal of Experimental Social Psychology*, *43*(3), 489–496. doi:10.1016/j.jesp.2006.05.003

de-Marcos, L., Garcia-Lopez, E., & Garcia-Cabot, A. (2016). On the effectiveness of game-like and social approaches in learning: Comparing educational gaming, gamification & social networking. *Computers & Education*, *95*, 99–113. doi:10.1016/j.compedu.2015.12.008

Ding, X., Yue, X., Zheng, R., Bi, C., Li, D., & Yao, G. (2019). Classifying major depression patients and healthy controls using EEG, eye tracking and galvanic skin response data. *Journal of Affective Disorders*, *251*, 156–161. doi:10.1016/j.jad.2019.03.058 PMID:30925266

ElProCus. (2021). DHT11 Sensor Definition, Working and Applications. *ElProCus - Electronic Projects for Engineering Students*. Available: https://www.elprocus.com/a-brief-on-dht11-sensor/

Gábana Arellano, D., Tokarchuk, L., & Gunes, H. (2016). *Measuring Affective, Physiological and Behavioural Differences in Solo, Competitive and Collaborative Games*. Lecture Notes of the Institute for Computer Sciences, Social Informatics and Telecommunications Engineering. doi:10.1007/978-3-319-49616-0_18

Handri, S., Yajima, K., Nomura, S., Ogawa, N., Kurosawa, Y., & Fukumura, Y. (2010). Evaluation of Student's Physiological Response Towards E-Learning Courses Material by Using GSR Sensor. *2010 IEEE/ACIS 9th International Conference on Computer and Information Science*. 10.1109/ICIS.2010.92

Huizenga, W., Admiraal, W., Dam, G., & Voogt, J. (2019). Mobile game-based learning in secondary education: Students' immersion, game activities, team performance and learning outcomes. *Computers in Human Behavior, 99*, 137–143. doi:10.1016/j.chb.2019.05.020

Hwang, G., & Chang, S. (2015). Effects of a peer competition-based mobile learning approach on students' affective domain exhibition in social studies courses. *British Journal of Educational Technology, 47*(6), 1217–1231. doi:10.1111/bjet.12303

imotions. (2021). *Galvanic Skin Response (GSR): The Complete Pocket Guide*. Available: https://imotions.com/blog/galvanic-skin-response/

Jagušt, T., Botički, I., & So, H. (2018). Examining competitive, collaborative and adaptive gamification in young learners' math learning. *Computers & Education, 125*, 444–457. doi:10.1016/j.compedu.2018.06.022

Maulsby, R., & Edelberg, R. (1960). The interrelationship between the galvanic skin response, basal resistance, and temperature. *Journal of Comparative and Physiological Psychology, 53*(5), 475–479. doi:10.1037/h0041916 PMID:13768388

Naveen Kumar, C., & Shivakumar, G. (2018). *A Real Time Human Emotion Recognition System Using Respiration Parameters and ECG*. Intelligent Human Computer Interaction. doi:10.1007/978-3-030-04021-5_4

Neulog. (2021). Available: https://neulog.com/Downloads/NeuLog_User_Guide_Ver_6_11b.pdf

Nozawa, A., & Uchida, M. (2009). *Characterization of preference for viscosity and fragrance of cosmetic emulsions by autonomous nervous system activity*. ICCAS-SICE.

Pandey, A. (2021). 6 Killer Examples Of Gamification In eLearning (Updated In 2020). *eLearning Industry*. Available: https://elearningindustry.com/6-killer-examples-gamification-in-elearning

Prasolenko, O., Burko, D., Tolmachov, I., Gyulyev, N., Galkin, A., & Lobashov, O. (2019). Creating safer routing for urban freight transportation. *Transportation Research Procedia, 39*, 417–427. doi:10.1016/j.trpro.2019.06.044

Salomon, G. (1984). Television is "easy" and print is "tough": The differential investment of mental effort in learning as a function of perceptions and attributions. *Journal of Educational Psychology, 76*(4), 647–658. doi:10.1037/0022-0663.76.4.647

Scotti, S. (2006). Automatic Quantitative Evaluation of Emotions in E-learning Applications. *2006 International Conference of the IEEE Engineering in Medicine and Biology Society*. 10.1109/IEMBS.2006.260601

Sharma, A., Khosla, A., & Khosla, M. (2018). Efficacy of an Android-based Game Intervention in Enhancement of Face Recognition Skills for Children with Autism. *International Journal of Learning and Change, 1*(1), 1. doi:10.1504/IJLC.2018.10011657

Sharma, R., Goel, D., Srivastav, M., & Dhasmana, R. (2017). Psycho-Physiological Parameters of Nurses in Critical and Non-Critical Units. *International Journal of Nursing Sciences, 7*(5), 107–110.

Sonali, A. (2020). Effect of PUBG on Stress Level in Youngsters Using GSR Sensor. *UGC Care Journal, 40*, 1142–1149.

Tijs, T., Brokken, D., & IJsselsteijn, W. (2008). Creating an Emotionally Adaptive Game. *Lecture Notes in Computer Science, 5309*, 122–133. doi:10.1007/978-3-540-89222-9_14

Tobii Pro. (n.d.). Retrieved from https://www.tobiipro.com/learn-and-support/learn/GSR-essentials/how-does-a-gsr- sensor-work/

Troussas, C., Krouska, A., & Sgouropoulou, C. (2020). Collaboration and fuzzy-modeled personalization for mobile game-based learning in higher education. *Computers & Education, 144*, 103698. doi:10.1016/j.compedu.2019.103698

Udovičić, G., Đerek, J., Russo, M., & Sikora, M. (2017). Wearable Emotion Recognition System based on GSR and PPG Signals. *Proceedings of the 2nd International Workshop on Multimedia for Personal Health and Health Care*. 10.1145/3132635.3132641

Urh, M., Vukovic, G., Jereb, E., & Pintar, R. (2015). The Model for Introduction of Gamification into E-learning in Higher Education. *Procedia: Social and Behavioral Sciences, 197*, 388–397. doi:10.1016/j.sbspro.2015.07.154

Villagrasa, S., Fonseca, D., Redondo, E., & Duran, J. (2014). Teaching Case of Gamification and Visual Technologies for Education. *Journal of Cases on Information Technology*, *16*(4), 38–57. doi:10.4018/jcit.2014100104

Villarejo, M., Zapirain, B., & Zorrilla, A. (2012). A Stress Sensor Based on Galvanic Skin Response (GSR) Controlled by ZigBee. *Sensors (Basel)*, *12*(5), 6075–6101. doi:10.3390120506075 PMID:22778631

Wang, C., Baird, T., Huang, J., Coutinho, J., Brien, D., & Munoz, D. (2018). Arousal Effects on Pupil Size, Heart Rate, and Skin Conductance in an Emotional Face Task. *Frontiers in Neurology*, *9*, 1029. Advance online publication. doi:10.3389/fneur.2018.01029 PMID:30559707

Wang, W., Weng, J.-F., Su, J.-M., & Tseng, S.-S. (2004). Learning portfolio analysis and mining in SCORM compliant environment. *34th Annual Frontiers in Education. FIE, 2004*, 180–187. Advance online publication. doi:10.1109/FIE.2004.1408490

Werbach, K., & Hunter, D. (2012). For the Win: How game thinking can revolutionize your business. doi:10.1017/CBO9781107415324.004

Yildirim, I. (2017). The effects of gamification-based teaching practices on student achievement and students' attitudes toward lessons. *The Internet and Higher Education*, *33*, 86–92. doi:10.1016/j.iheduc.2017.02.002

# Chapter 3
# Latest Advancements in Wearable Devices:
## A Review

**Mohit Angurala**
*Chitkara University Institute of Engineering and Technology, Chitkara University, Punjab, India*

## ABSTRACT

*Wearable devices have impacted the daily life of every individual. These devices come with the embedded feature that fits almost within clothes, accessories, or even watches. One of the wearables named "wrist-worn devices" has gained acceptance by the masses among other wearables. Vital information can be easily gathered with the help of such wearables. These are also suitable for myriad applications such as sports, agriculture, medical, and several more. This chapter gives a comprehensive review of wearable computing electronic devices being used in various fields and provides the latest trends in wearables. The main objective of this review is to discuss various challenges that are faced by individuals in using wearables and the latest methods that can overcome these issues. In the modern epoch, wireless wearable devices have gained a lot of importance in various fields such as sports, agriculture, medical, and many more.*

## INTRODUCTION

Today's modern world have ignited an innovative human-computer interaction system with latest wearables which led to the quick expansion in the communication technology. It further facilitates connectivity amid users so that they can access the

DOI: 10.4018/978-1-7998-7433-1.ch003

information online easily. Wearable technologies (WTs) are cheap and affordable electronic devices that obviates the requirement of healthcare professionals and facilities. Thus, these WT are optimum sources for delivering mobile healthcare services to individuals (Alhaddi, 2016). The cell phones nowadays are quite advanced as they are equipped with lot of features such as: high-resolution cameras, data processing capabilities, accelerometer, fingerprinting, and many more. Also, they share excellent connectivity with cloud so that they can collect and store personalized data, which can be assessed by certified doctors remotely. These practices are commonly used in telemedicine applications, particularly when nations are combating against emergency situations such as covid-19. WTs are networked devices that particularly gathers data and track activities. These WTs are embedded with miniature small sensors (chips) mounted into consumer electronics, accessories and attires (Bigger, 2016). The used sensors in WTs measures humidity, temperature, heartbeat and many more. Wearable devices are used in different application areas such as: healthcare, agriculture, sports etc. Each wearable device is equipped with sensors, processor, network ports cameras to permit monitoring and synchronizing multiple parameters. Wearables are lightweight, modest shape and multifunctional that perform the task of sensing, analyzing, storing and transmitting the data. However, the processing of data may occur either on the wearer or at a remote location. One best example of this is the release of dangerous gases alert when it is detected (Colley, et al, 2016).

## BACKGROUND

WTs should meet the expectations of normal clothing or in other words, the wearable devices must be soft, flexible and washable. A smart attire ideally consists of sensors, electronic components and power, but in reality, there are certain limitations with these components (Colley, et al, 2016). The wearable computers in the past were too bulky, but with an increased miniaturization of electronic devices, the WTs are becoming more integrated. By creating a smart sensor based fabric, the attires can retain its normal tangible features. The design totally depends on the need of user and the application. Wearables were firstly developed in the 13th century when eyeglasses became available (Benedetto, et al, 2017). Then, after 300 years, the first wearable clocks were invented in the form of wristwatches. Further, after 300 years again the rise of the first wearable calculators, music systems and computers were invented Braun, et al, 2006).

In 2002, Bluetooth technology gained a lot of market that attracted thousands of customers across the globe which allowed wireless interactions with the devices. In this era, fitness/smartwatch devices dominate the market. Wearables are expected to dominate the fitness market. Wearable devices that track the progress will continue to hold popularity (Global Organic Textile Standard, 2019). Healthcare is another sector that looks set to be improved by new wearables. Healthcare organizations are investing huge amounts of money on latest technologies which will bring developments to the healthcare systems in the next few years.

## MAIN FOCUS OF THE CHAPTER

The main aim of this chapter is not only limited to find out the latest trends in the wearable devices, but also to explore the various application areas of WTs. This chapter also look out for possible solutions toward the challenges being faced in wearable Internet of Things (IOTs). The next sub-section discusses different application areas of wearable devices.

### Application Areas of Wearable Devices

There are different kinds of wearable devices that are gaining the attention in the market today. Some of the application areas of wearable devices are (Hunt, 2018):

**Improved Productivity Level:** There are many organizations which are offering wearable devices to their workers. The data gathered using these wearables allow organizations to gauge their approaches in a way to assist team members in boosting their productivity.

**Diet Management:** Over-weight individuals may take benefit of health aware wearable devices, that makes use of an in-body sensor to successfully measure vital nutritional parameters and comparing it with cloud-based applications to provide the user with the information in the form of valuable feedbacks on what diet is essential for them to make them fit.

**Saving Lives:** Wearable devices can help relieve pain, ensure road safety, change globetrotting and curb accidents.

**One Ring to Rule All:** Myriad of companies has designed smart rings which allows individuals to use it during shopping. Using the smart ring, one can get access to their car or turn on smart home appliances.

**Sexual Assault:** Some wearables are manufactured to send alerts to the parents or police officers whenever it detects force and falls, via Bluetooth, thus helping women which further reduces crimes.

**Biosensors:** The wearable biosensors are patches which collects factual data on their body, such as: heart rate, temperature and movements.

Some of the industrial applications of wearable devices are shown in table 1 (Hunt, 2018, Global Organic Textile Standard, 2019, Colley, 2016).

*Table 1. Industrial applications*

| Sr No. | Industry | Applications |
|---|---|---|
| 1 | Healthcare and Medical | Blood pressure monitors, Electrocardiogram monitors, Defibrilators, Drug delivery products, Insulin pumps and Smart glasses |
| 2 | Industrial | Smart Personal Protective Equipment (PPE), Smart watches and glasses, Exoskeletons and Body-worn terminal |
| 3 | Defense | Exoskeletons, Smart tactical vest, Body-worn terminals |
| 4 | Sports | Activity monitors, Fitness & heart rate monitors, Foot pods & pedometers, Sleep sensors, Smart glasses, clothing and watches |

## Issues in Wearables

The biggest challenge for the wearable industry is to get a sustainable customer engagement. Myriad of electronic wearables are short lived because of

- Temporary engagement of customers
- Poor Quality
- Difficult to sync with smartphones
- Short battery life
- Poor design and many more

However, individuals may find WTs to be robust (functionally and physically) due to which they are more attracted towards such devices and find it useful for them Hischier, 2015, Kasser, 1999, Bigger, 2016).

# SOLUTIONS AND RECOMMENDATIONS

Even though the WTs produce innovative and unique challenges, yet there are numerous prospects that must be apprehended only if wearable devices are produced in sustainable manner. Few latest trends which emerged as solutions toward overcoming the issues of WTs are as follows (Köhler, 2004, Bigger, 2016, Hunt, 2018, Global Organic Textile Standard, 2019):

## Material Selection

True sustainable production needs product that must be produced in such a manner that the valuable resources inside it are not lost. The biodegradable materials must be used so that they can be securely returned to the environment.

## Manufacturing

In terms of WTs, different toolkits have been emerged- which are helping both the designers and end-users to develop functional prototype of wearables. One of the examples of wearable prototypes is Lilypad, which is Arduino platform based and enables the end-users to create wearables. Adafruit also developed similar wearable electronics components called flora.

## Distribution

Distribution of products to the market outcomes in substantial impacts to the environment. Designers can plan the delivery stations/channels to reduce this intangible element.

## Use and Consumption

Smart textiles are often introduced as a possible expansion that could render the textile industry more sustainable as they are anticipated to improve textile toughness to alter the way individuals use to handle traditional way of handling attires. Self-functioning textiles is one of the innovative methods which has the abilities of photocatalytic self-cleaning coating that removes bacteria/strains and odor by degrading dirt particles.

## Disposal

The amount of e-waste that is considered a sustainability problem, is likely to increase when wearables reach their end of life and give rise to new disposal and recycling issues. In this respect, Product Service Systems (PSSs) could be beneficial because they encourage the consumer to return pieces in exchange for another. This increases rates of recycling and ensuring a reduction in disposal, thus reducing e-waste.

# FUTURE RESEARCH DIRECTIONS

The future of WTs is continuously improving as machine learning algorithms are permitting the conversion of sensor input into activity data. As the size of Integrated circuits (ICs) are getting smaller, the way of wearables usage is getting far better. Therefore, the improvements at both algorithmic level and design level of wearables could attract the market more effectively (Hunt, 2018, Di Benedetto, 2017).

## CONCLUSION

In this chapter, a detailed analysis of literature across the fashion industry for WTs to understand the challenges and progress in achieving environmental sustainability aims is performed for the product development life cycle. The chapter also illustrates the methods to address the challenges in WTs. Further, various applications of WTs and its associated issues and latest trends to overcome the problems in WTs areas are also discussed.

## FUNDING INFORMATION

This research received no specific grant from any funding agency in the public, commercial, or not-for-profit sectors.

## REFERENCES

Alhaddi, H. (2015). Triple bottom line and sustainability: A literature review. *Bus. Manag. Stud.*, *1*(2), 6–10. doi:10.11114/bms.v1i2.752

Bigger, E., & Fraguada, L. (2016). Programmable Plaid: The Search for Seamless Integration in Fashion and Technology. *Proceedings of the 2016 ACM International Joint Conference on Pervasive and Ubiquitous Computing(UbiComp'16)*, 464–46. 10.1145/2968219.2971343

Braun, V., & Clarke, V. (2006). Using thematic analysis in psychology. *Qualitative Research in Psychology*, *3*(2), 77–101. doi:10.1191/1478088706qp063oa

Bryman, A. (2012). *Social Research Methods* (4th ed.). Oxford University Press.

Colley, A., Pakanen, M., Koskinen, S., Mikkonen, K., & Häkkilä, J. (2016). Smart Handbag as a Wearable PublicDisplay–Exploring Concepts and User Perceptions. *Proceedings of the 7th Augmented Human International Conference (AH'16)*, 1.

Di Benedetto, A. C. (2017). Corporate social responsibility as emerging business model in fashion marketing. *J. Glob.Fash. Mark.*, *8*(4), 251–265. doi:10.1080/20932685.2017.1329023

GlobalF. A. T. (2020). Available online: https://globalfashionagenda.com/commitment/#toolboxes

GlobalO. T. S. (2019). Available online: https://www.global-standard.org/the-standard/general-description.html

Hischier, R., Coroama, V. C., Schien, D., & Achachlouei, M. A. (2015). Grey energy and environmental impacts of ICThardware. In ICT Innovations for Sustainability. Springer.

Hunt, C. P., Wickham, M., Pittson, R., & Lewison, J. (2018). *Reusable, Unzippable, Sustainable Electronics (ReUse) Interconnect System for the Circular Economy*. Report for NPL, MAT 75. Available online: https://eprintspublications.npl.co.uk/6645/

Köhler, A., & Erdmann, L. (2004). Expected environmental impacts of pervasive computing. *Hum. Ecol. Risk Assess*, *10*, 831–852.

# Chapter 4
# Role of 5G Communication Along With Blockchain Security in Brain-Computer Interfacing: A Review

**Mohammad Mudassir Ahmad**
*DAV Institute of Engineering and Technology, India*

**Kiran Ahuja**
https://orcid.org/0000-0002-1213-8010
*DAV Institute of Engineering and Technology, India*

## ABSTRACT

*The electroencephalogram is used in brain-computer interface (BCI) in which signal from the human brain is sensed with the help of EEG and then sent to the computer to control the external device without having any touch of muscular body parts. On the other hand, the brain chip interfacing (BCHIs) is a microelectronic chip that has physical connections with the neurons for the transfer of information. The BCI needs a reliable, high-speed network and new security tool that can assist BCI technology. 5G network and blockchain technology is ideal to support the growing needs of brain chip interfacing. Further, the Cloudmind, which is an emerging application of BCI, can be conceptualized by using blockchain technology. In this chapter, brain-computer interfaces (BCIs) are expedient to bridge the connectivity chasm between human and machine (computer) systems via 5G technologies, which offers minimal latency, faster speeds, and stronger bandwidth connectivity with strong cryptographic qualities of blockchain technologies.*

DOI: 10.4018/978-1-7998-7433-1.ch004

## INTRODUCTION

The brain-computer interface (BCI) is a technology that allows controlling the machines with the help of human brain signals. To control the external device, The BCI uses the following steps. (i) Collecting the brain signal. (ii) Compiling the signal into a machine-readable form. (iii) Controlling the machine according to the signal received from the human brain. The brain-computer interface is used in various medical applications such as in replacing the lost motor functionality of paralyzed patients, neurofeedback, rehabilitating the moving ability for physically locked-in patients. The main objective of BCI is to enable or restore the important functions of the patients affected by several neuromuscular disorders such as amyotrophic lateral sclerosis, cerebral palsy. From the initial demonstration of the single neuron-based control system, the research community and the scientists are now able to control the complex movements of a robotic arm, wheelchair, cursor, prostheses, and other devices by using the electroencephalographic, intracortical, electrocorticographic, and other brain signals. Due to advancements in microelectronic technology and nanotechnology, the application of the microelectromechanical system (MEMS) and neural nanorobotics has been increased in recent years. Due to this massive advancement, the MEMS and nanorobotics have been now accepted and used by many neuroscientists for the recording of living neurons.

The brain chip interfacing is classified into three categories based on biological entities involved namely; neurons, tissue, and brain by Fromherz, P. (2003). In the first level of interfacing, the neurons are linked with the microelectrodes or electrolyte-oxide-semiconductor-capacitors. these microelectrodes measure and record the electrical signal of the neurons given by W.L. Rutten Annu (2002), Potter et al. (2006). The brainstorm project is a recent example of brain chip interfacing (BCHIs) in which tight electrical coupling between microchip and neurons is done with the help of a gold micro nail shaped microelectrode that is surrounded by neurons with the help of the phagocytosis like method by A. Hai et al. (2010). In the second level of interfacing, the concept of establishing the interaction between the brain tissues and microchip is used by placing a slice of tissues on the chip. Finally, in the third level of interfacing, the microchip is inserted into the human brain or in the other part such as the spinal cord or sensory organs with the help of surgery suggested by S. Girardi et al. (2010).

Further, the neural nanorobots may be used to interface the human brain with a computer cloud system. This is called Brain/Cloud interface. The nanoparticles are injected in the human brain through the vasculature in a proper clinical environment and after passing through the blood-brain-barrier, these nanorobots position themselves with the brain cells. Generally, the neocortex of the human brain is used as a gate by the neural nanorobots to communicate with external cloud supercomputer. As mentioned, the BCI or neural nanorobot communicates the brain signal to the external machine so that an appropriate task can be performed accordingly.

The transfer of the signals from the human brain to the machine or vice versa required high speed, wireless, and reliable network. Currently available wireless cellular network (4G LTE) is not capable of fulfilling the growing need for BCI applications. The currently available 4G network can provide speed up to 100 Mbits/sec and the reliability of the network is not that much good. There is a need for a new network interface which is reliable as well as offers high speed. 5G or Fifth generation cellular network can act as a leading wireless communication link between the human brain and external machine. The upcoming 5G technology has features that can support BCI technology. 5G network interface has high speed, high reliability and it is much more secure as compare to the 4G stated by Martins et al. (2019).

However, the security of the BCI technology not only depends upon the network interface but some other factors as well. In BCI, the brain signal is first converted into a machine-readable signal and then this machine-readable signal is sent as a command to the device/machine. Due to this, BCI technology is very prone to data manipulation by some external body, and therefore due to this insecurity, BCI technology can be easily hacked. Here comes the role of Blockchain technology. The blockchain platform provides the necessary data security to the BCI user and prevents attacks such as sniffing, eavesdropping, and hijacking. One of the applications where both BCI technology and Blockchain technology can be implemented together is Cloudmind or Crowdmind. The practical view of implementation is presented in Fig. 1.

*Figure 1. Represented a 5G facilitated communication between BCIs along with blockchain encryption*

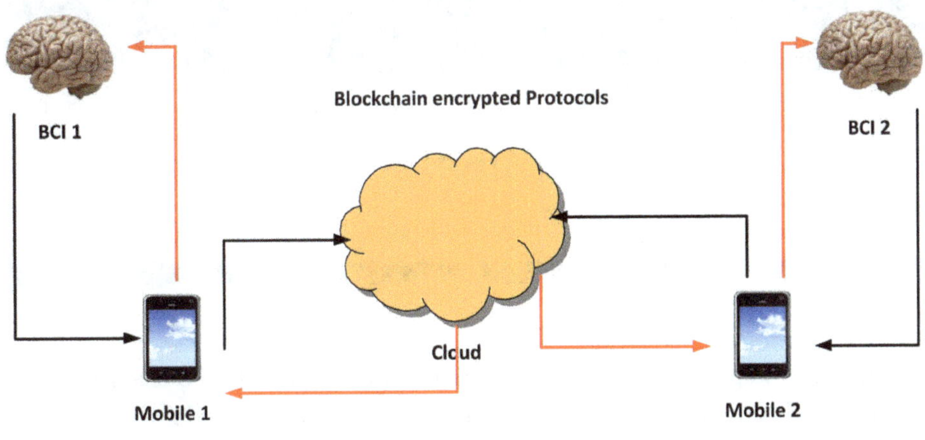

BCI integrated human body will be communicated signal to a portable mobile station (cell phone gadget acts as a transducer and processor). The received signals will be processed, transduced and self-sufficiently sent to the cloud. Data demands and showing up signals can likewise to be prepared and transduced from the cloud of the mobile stations and send to the beneficiary BCI continuously. Communication encouraged by the cloud may require the formation of a hyper, scrambled Brain Over Internet Protocol (BoIP) separate from our current protocols. The inherent, solid cryptographic characteristics of blockchain advancements may help in guaranteeing that these BoIP networks are can't be easily invaded. The arrangement would allow for liberated data openness (transfers and downloads to and from human subjects just as among people and machines). People in reverberation with one another or with machines (synchronization of figured waves) would have the option to convey and get to data at unrestrained choice.

Here, Blockchain technology stores the data in the form of blocks. These blocks are interconnected with each other in the form of chains. The blockchain mainly consists of three concepts: Blocks, Miners, and Nodes as shown in Fig. 2. The block in a blockchain further has three elements namely: Data, Nonce and Hash. The data or information for a particular task is stored in this section of the block. The data may be records of the user or complex command sent to the machine in BCI. Whereas, an nonce is a 32-bit number that generated randomly whenever a block is created in the blockchain. In addition, the hash is generated by a nonce. It is a 256-bit number and generally starts with zeroes. Further whenever the block is created or added in the chain, the nonce and hash tied forever with the data of the block until mining is done.

*Figure 2. Basic working concepts of blockchain technology*

Whereas, the process of creating or adding a new block in the chain is called mining. As already discussed, every block has its unique hash and nonce. Further, each hash in chain references the previous hash so it is very difficult to add or create a new block in a chain. For adding a block in the chain, it is mandatory to find a nonce with an acceptable hash. As the nonce is a 32-bit number and hash is a 256-bit number so, there are around 4 billion hash-nonce combinations before the right one found. The nodes allowed the decentralized feature in the blockchain technology. Nodes can be any electronic device that keeps the copies of the blockchain and each node has its copy of the blockchain (BuiltIn, n.d.).

**Organization of the Chapter:** This chapter survey's the perfect relation between BCIs', 5G communication requirement along with blockchain security. Section 2 highlights the literature survey giving the current standards related to BCIs'. Section 3 described the diverse types of BCI interfacing available in the market. Section 4 provides the details of BCIs application in diverse medical applications. Role of 5G in BCI technology is addressed in Section 5. Whereas, Section 6, shows glimpse of blockchain technology implementation in medical field. Lastly, Section 7 concludes this chapter with the final reconsiderations.

## RESEARCH AND APPROACH

Hans Berger's contribution to the brain-computer interfacing is very significant. In 1924, Hans Berger developed the first device that measures the signal of the brain called EEG. Richard canon's discovery of the electrical signal in the animal's brain in 1875 was an inspiration for Hans Berger. In 1959, the concept of nanotechnology introduced by the physicist Richard Feynman in his speech "There is Plenty of Room at the Bottom". In 1970, the University of California started research on brain-computer interfacing. The main aim of the research was to restore the functionality of the motor and restore the movement of paralyzed patients. In 1986, K. Eric Drexler put the idea of nanomachine on the map with his book "Engines of Creation: The Coming Era of Nanotechnology" cited by Diamandis et al. (2016). In 1998, the researcher Philip Kennedy implanted the first brain-computer interface object in the human brain. However, the function of the object as limited. In 2001, John Donogue and his team from Brown University establish a company name cyber kinetics. The neuro Port was the first product, the company came up with. In 2002, the two monkeys were fitted with microelectrodes or microchips (Study: Monkey brains move cursor, 2002). In June 2004, there was a significant development in the field of BCI when a man name Mathew Nagel became the first person that was implanted with BCI. The company name Ambient in 2008 demonstrates the world's first voiceless phone call in the Texas Instrument Developer Conference (TIDC).

Audio then transfers this signal to the phone connected through Bluetooth cited by Diamandis et al. (2016). In the year 2014, the researcher from the University of

Washington successfully demonstrates the brain to brain communication through the internet. The second participant was hooked with a magnetic stimulation coil which controls the movement of the hand Martone et al. (2019). Recently, the scientists of EPFL (Ecole Polytechnique Fédérale de Lausanne) tested a new neuroprosthetics technology that modulates the user's voluntary control with the robotic control. For now, the testing of the algorithm is claimed by Ziegler-Graham et al. (2008), Wei, Li et al. (2010), Tan et al. (2010), Kim et al. (2011), Vecchiato et al. (2011), Siang Ang et al. (2011), Lin et al. (2012), Švogor et al. (2012), Birbaumer et al. (2013), Ruiz et al. (2014).

When coming to the blockchain technology, in 1982 David Chaum proposed a blockchain-like protocol in his paper" Computer Systems Established, Maintained, and Trusted by Mutually Suspicious Groups". In 1991, Stuart Haber and W. Scott Stornetta further work on the secured chain of the blocks. Haber, Stornetta, and Dave Bayer in 1992 proposed an improved Markle Tree design that increases the efficiency by collecting the document certificate into one block. In 2008, the first blockchain was realized by Satoshi Nakamoto. Nakamoto improved the design by using the hash-cash method that timestamp the block without the need of any third party or trusted party. This design was used in the cryptocurrency bitcoins. In 2014, the blockchain file size of the bitcoin that contains all the transaction records almost reached 20 GB and in 2015 the blockchain size for the transaction reached 30 GB. In 2017, the size of the blockchain file touch 100GB. In early 2020, the blockchain file size already exceeds 200GB (Blockchain History, n.d.).

Recently some startup has joined the field such as Facebook, Kernel, Neuralink. In 2017, Facebook started hiring neuroscientists and BCI engineer for their project named B8. The main aim is to detect the word of the Facebook user and allow the users to type the message only by thinking. This allows the user to type 100 words per minute. For this project, Facebook is collaborating with the Washington University school of medicine, UC San Francisco, John Hopkins Medicine, UC Berkeley, John Hopkins University's Applied Physics Laboratory. It is working on the software and hardware to build a device that converts the sound into the specific brain frequency. The company Neuralink was established by the Elon Musk in 2017. The main of the company is to build an implantable device that helps humans to keep uprise with artificial intelligence. In the beginning, the implantable device could be used to detect diseases such as brain tumors, depression, etc. The main advantage of the implantable device is that it is due to this the signal quality is better than EEG. However, invasive technology has some drawbacks such as battery life, the degradation of the signal with time. The CEO of the company said that it may take around four to five years for the meaningful implementation of the BCI. The main aim of the company is also to enhance human capabilities by using implantable devices. Bryan Johnson, the founder of the company invested around 100 million dollars to develop the brain implant. While keeping in view the importance of BCIs' in current scenario the details of its interfacing as well as its application in medical field described in following sections. Importance of 5G communications and need of blockchain security in also elaborated for enhancing the wider vision in this field.

## TYPES OF BRAIN-COMPUTER INTERFACING

The BCI system can be divided into three depending upon the methods (i.e. Non-invasive, Semi-invasive and Invasive) of measurement of the neuron signals. The details of BCI system are presented in Fig. 3.

### Non-Invasive

In the Non-invasive, the sensor which measures the electrical or magnetic signals produced by the brain is placed outside the scalp. The techniques used in the non-invasive are as follows:

**EEG (Electroencephalography):** The EEG is the most common technique used for the measurement of the signals produced by the brain. In the human brain, there is electric current produce by the neurons. The EEG measures this electric current from the scalp to measure the brain signal.

**MRI (Magnetic Resonance Imaging):** MRI is a technique in which a very high magnetic field interacts with the protons present in the human body to produce an image or neural image. When the human body interacts with a high magnetic field, all the free hydrogen nuclei present in the body align themselves by the respective magnetic field. This effect is called Larmor Precession. During the alignment process, the nuclei emit some energy which is captured by the machine. Based on this energy, an appropriate 3D gray image is formed said by Brown et al. (2011).

**fMRI (Functional Magnetic Resonance Imaging):** fMRI is a procedure of taking neuroimaging of the human brain. The fMRI is based on the fact that the blood flow and the human brain neural activity are linked together said by Huettel et al. (2004). In the human brain, the hemoglobin transfers the oxygen to the neurons. When, the neurons activated the consumption of the oxygen increases, which resulted in an increase in the blood flow. The magnetic character of the hemoglobin changes depending upon the amount of the oxygen present in it which can be detected with the help of an MRI machine which is a powerful electromagnet.

*Figure 3. Hierarchy of BCI system*

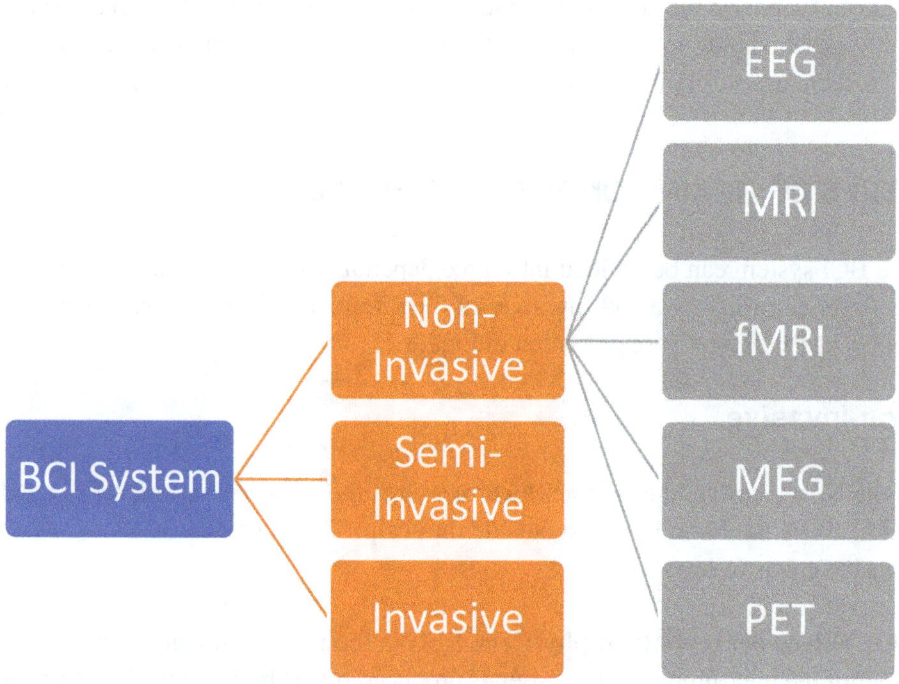


**MEG (Magnetoencephalography):** In this technique, the sensor is placed outside the scalp which determines the magnetic field produce by the electric current in the human brain. The resolution of the MEG is better than EEG because as MEG uses the magnetic field for the measurement of brain signal, there is less signal attenuation due the scalp and fluids in the brain which increases the signal to noise ratio of the receiving signal said by Lal et al. (2005).

**PET (Positron Emission Tomography):** PET is a nuclear imaging technique. In the PET, a small amount of radioactive element i.e. a nuclear tracer is introduced in the blood so that the nuclear tracer can reach the brain. In the brain, this nuclear tracer combines with the glucose to form radionuclide. The PET is used to observe neuro transmittance, metabolism, and some other processes claimed by Mesgarani et al. (2012).

## Semi-Invasive

In this technique, the electrodes are positioned inside the scalp to measure the brain signal. The semi-invasive uses the ECoG technique to measure the signals. In the ECoG or electrocorticography, the electrodes which measure the electrical signals are placed on the dura or arachnoid layer of the brain Waldert et al. (2016). Since the surgery is required to insert the electrodes in the brain. Therefore, this method is used only when the surgery is necessary or due to some other medical reasons. As the electrodes are inserted inside the scalp, there is less attenuation of the signal which increases its spatial resolution. The different layers of scalp are presented in Fig. 4.

*Figure 4. Different layers of the human scalp Abdullah-Al-Mamun et al. (2013)*

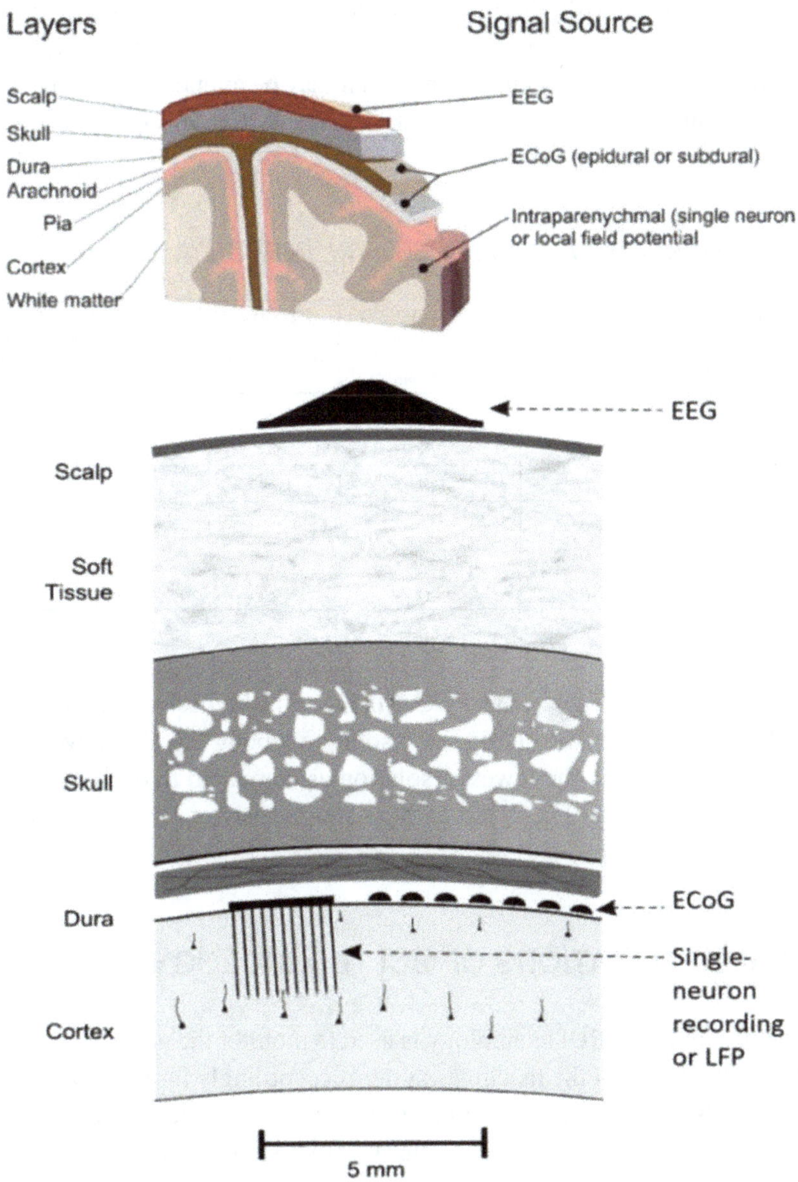

## Invasive

In this technique, the microelectrodes are directly inserted into the cortex of the human brain with the help of neurosurgery. The microelectrodes inserted can measure signals from a single area of the brain cell or multiple areas of the brain cell depending upon the requirement Abdulkader et al. (2015). The signal to noise ratio in this technique is very high as there is no attenuation due to the scalp and other fluids. However, the insertion of the microelectrodes requires neurosurgery which is expensive and risky. The body may not accept the microelectrode inserted inside the scalp and treat it as an external entity which may cause the building of the scar around the electrodes and therefore signal may be attenuated which decreases its quality. Comparative study between EEG, MRI, fMRI, MEG, PET is presented in Table 1.

*Table 1. Comparative study between EEG, MRI, fMRI, MEG, and PET*

| Parameters | EEG | MRI | fMRI | MEG | PET |
|---|---|---|---|---|---|
| Biological process measure | Neuro electrical signal | Nuclear magnetic resonance response. | Haemodynamic Response | Neuromagnetic signals | Hemodynamic response |
| Radiation | None | None | None | None | Yes |
| Invasive | None | None | None | None | None |
| Spatial-resolution | Low | High | High | High | High |
| Temporal-resolution | High | Low | Medium | High | Low |
| Measurement mechanism | Direct | Only the structure | Indirect | Direct | Indirect |
| Cost | Low | High | High | Medium | High |

It is concluded from Table 1 that EEG, MEG has the highest temporal resolution, and MRI, PET has the lowest temporal resolution among the all. However, in the case of Spatial, EEG has the lowest resolution as compared to other non-invasive techniques. Further, the EEG is the cheapest noninvasive techniques available at the present.

## MEDICAL APPLICATIONS OF BCI TECHNOLOGY

In the medical domain, BCI technology is used to monitor the neural activity of the human brain and control the machines/systems accordingly for paralyzed patients.

**Assistive System for Paralyzed Patients:** The BCI technology can be used to assist paralyzed patients or patients suffering from spinal cord injury. Any above mentioned BCI invasive or non- invasive techniques (EEG, MRI, etc.) can be used to record or monitor the neural activity of the brain and depending upon the neural activity the BCI system converts the brain signal into a machine-readable signal so that appropriate machine or device can be controlled with brain thoughts.

However, the quality and the strength of the signal obtained from the human brain depend upon the technique used to collect the brain signal. Generally, the neural signal obtains by using the invasive technique is noise-free as compare to the signal obtained with the help of non-invasive technique. This is because in invasive technique the electrodes are directly placed inside the human skull. Therefore, there is no distortion in the invasive signal that may be caused by skull and fluids in non-invasive techniques Nijboer et al. (2008).

**Disease Diagnosis and Detection:** The diseases such as brain tumors, sleep disorders, and brain swelling can be detected with EEG technique, MRI, or CT scan. The BCI chip placed in the human brain can regularly monitor neural activity. The BCI chip diagnosis the mental disorder based on brain neural activity. If there is any mental disease, it can be found in the early stage and appropriate treatment can be started for the same said by Zhang et al. (2019).

**Neuroprosthetics:** Neuroprosthetics is an emerging field in which the biological part of the body that loses functionality is replaced with an artificial part so that functionality can be restored. Neuroprosthetic can use BCI technology to acquire the brain signal and then use this brain signal to control the functionality of the artificial part. Neuroprosthetics can be used to replace a damaged arm that lost functionality with an artificial arm. Neuroprosthetics can be used to restore the visual sight of a blind patient. The external image acquiring device such as a camera can capture an image. This image signal is first converted into a brain signal and then sends to the visual cortex so that the blind patient can see objects even when no light can enter into his/her actual damaged eye claimed by Horch et al. (2017), Hazubski et al. (2019).

## ROLE OF 5G IN BRAIN CHIP INTERFACING

Every application of BCI we have discussed so far requires a high-speed wireless communication network. The question may be arising why we need wireless communication in BCI as the transfer of signal from the brain to the computer can take place with a physical medium such as optical fiber. The answer is that the wireless network provides more mobility and free movement to the patient or user of BCI as compared to wired/physical medium. Brain/Cloud Interface that allows the human brain to communicate with external computers also needed a very reliable communication network.

Currently available cellular network(4G) provides a maximum speed of 100 Mbits/sec which is not sufficient to fulfill the need of BCI. Further, internet traffic will reach 30 GB per capita by 2021. Therefore, the bandwidth and speed offer by the current 4G network is not sufficient for the implementation of BCI. Therefore, there is a need for new high speed and reliable network. 5G technology is a fifth-generation cellular network that is much faster and reliable as compared to the previous 4G technology. 5G offers a speed up to 10 Gigabits/second which is 600 times faster than 4G. Further, 1000x bandwidth per unit area and 1-millisecond latency of 5G makes it an ideal wireless network for brain chip interfacing.

## Features of 5G

There are some features and specification of 5G that make it better than any previously available cellular network. The pictorial view of 5G features is shown in Fig. 5.

*Figure 5. Benefits of 5G technology*

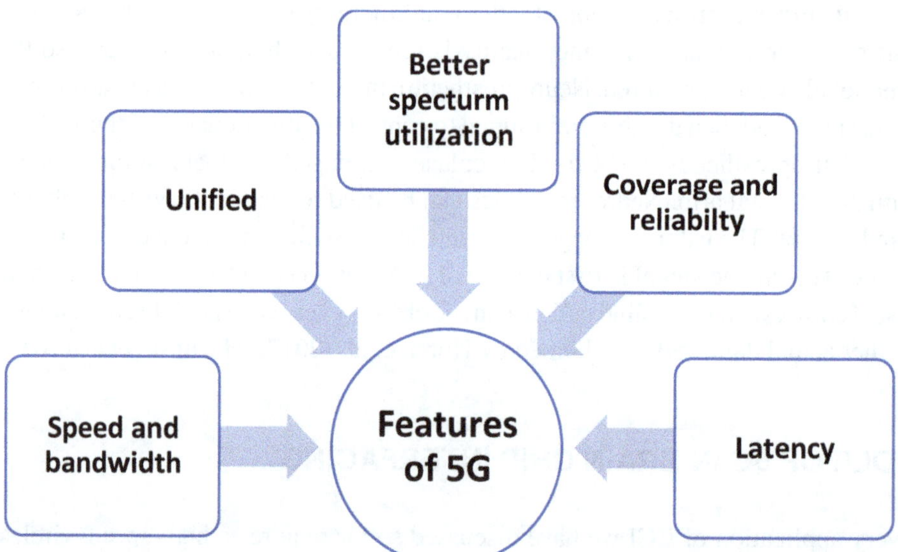


**Speed and Bandwidth:** 5G is significantly faster than any other cellular network (like 1G, 2G, 3G, 4G). As already mentioned, 5G offers 600 times more speed as compared to 4G. Further, as already mentioned 5G can provide bandwidth up to 1000x per unit area.

**Unified:** 5G is a more unified platform as compare to 4G and therefore along with mobile broadband, 5G can be used in other applications such as in IoT, critical mission communication, and more.

**Better Spectrum Utilization:** 5G uses the available spectrum in a very efficient way so that no wastage of spectrum occurs. Due to this, both the speed and bandwidth of 5G increases significantly.

**Coverage and Reliability:** 5G offers coverage of 100% and availability of 99.9% which makes it a very reliable cellular network.

**Latency:** The time required by the data packet to be transmitted to the destination point and after that acknowledgment signal to reach the source point is called latency. The higher the latency, the poor is the network. The lower the latency, the better is the network. As already mentioned, the 5G network has a latency of 1 millisecond which is very good in terms of network performance by Obraczka et al. (2000), Shien et al. (2019).

## 5G Architecture

After going through the features of 5G, it is important to understand the architecture of 5G, which is presented in Fig. 6. The architecture is responsible for such type of specifications that 5G technology offers. Unlike previous cellular networks, 5G uses a more intelligent radio access network (RAN) architecture approach. The designing and specification of the 5G network are under 3GPP that is a 3rd generation partnership project. In 5G, multiple frequency ranges are given. The radio spectrum from frequency 30 GHz to 300 GHz is known as millimeter-wave and frequency range from 24 GHz to 100 GHz can be further used for multiple 5G applications. In millimeter-wave, the unutilized frequency ranges from 300 MHz to 3 GHz are reuse for the 5G network.

**MEC:** Multi-Access Edge Computing or MEC is a cloud-based computing technology that allows the applications stored in a central data center to move in system edge. The benefit of MEC is that the system edge is much closer to the end-user as compare to the centralized data center. Therefore, MEC creates a shortcut between the user and the central data center. If MEC is deployed in 5G then this reduces the latency and increases the bandwidth of the network.

**NFV in 5G:** The NFV or network function virtualization separates the software from the hardware and replaces the system functions such as firewalls and load balancers with virtualized software. The NFV eliminates the cost of expensive hardware as well as reduces the installation time required to install particular hardware. NFV allows virtualized network functionality in the 5G infrastructure. Any problem in the network can be detected easily with NFV and appropriate fast action can be taken accordingly.

**5G Core:** As already mentioned, the specifications and standards of the 5G network are defined by 3GPP. The new 5G network uses cloud aligned and service-based architecture. The 5G network further utilized the concept of virtual network functioning of NFV along with MEC architecture. These features and specifications of the 5G network make it able to support the increased throughput demand claimed by 5GPPP Architecture Working Group (2017).

*Figure 6. Basic preview of 5G architecture provided by 5GPPP architecture working group (2017).*

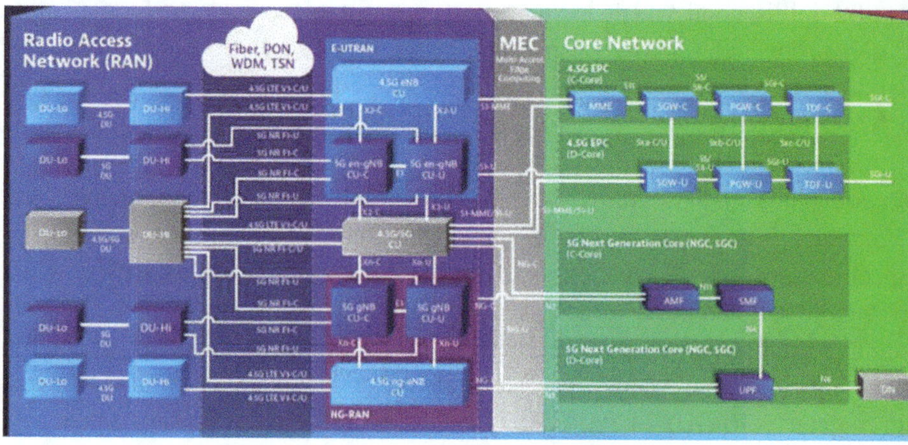


## BLOCKCHAIN TECHNOLOGY AND DATA SECURITY

As already mentioned, there are lots of signal/data transferring occurs between brain chip through the human brain to different peripherals devices (prosthetic arm, IoT devices, crowdmind/cloudmind, etc.). This makes BCI technology very prone to sniffing, eavesdropping, and hijacking. Currently, available data security measures are not enough to protect the BCI technology from hackers. There is a need for a better security system that can protect important data and information in BCI technology from stealing. Blockchain can be considered as an ideal tool that provides necessary security support to BCI technology. Blockchain is a system that records the data in a form such that recorded data cannot be manipulated or stolen by any third person. Blockchain is also called as Distributed Ledger Technology or DLT because it is a digital ledger of transactions that is distributed on the entire computer network given by Nofer et al. (2017). Whenever a transaction or transfer of signal occurs in the system then a 256-bits alpha-numeric hash and 32-bits nonce stored in the block along with the transaction data. Each block is connected with another block in form of a chain. Further, each hash of transaction is stored not only with the current transaction but also with the previous transaction and if there is any change in the transaction data, the hash associated with that transaction will change automatically. So, if anyone tries to change the transactional data then the person has to change all the hash associated with that transaction. That's why it's very difficult to change or manipulate data in the blockchain Gupta et al. (2017). The basic block diagram of working of blockchain technology is presented in Fig. 7. The logical components and types of blockchain implementation are described as follows:

*Figure 7. How the blockchain works provided by PriceWaterhouseCoopers (2016).*

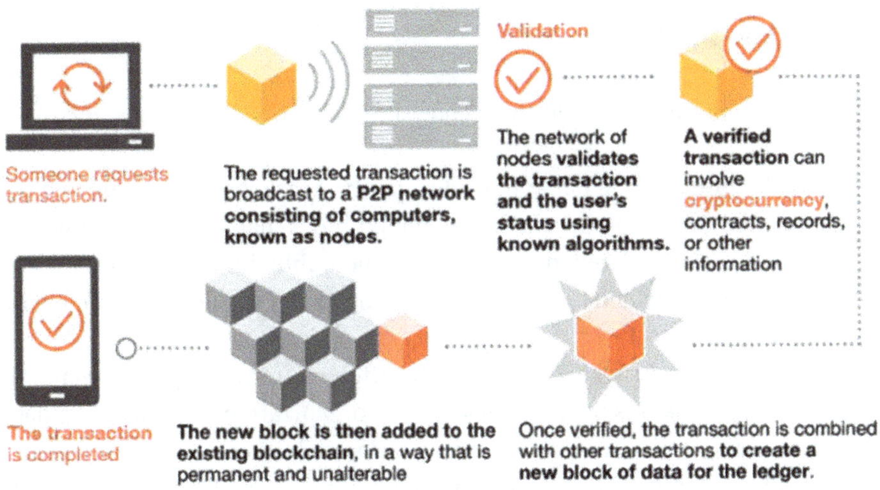


## Logical Components of the Blockchain

There are four important components of the blockchain.

**Node Application:** Node application is software that allows a particular computer or any other device to participate in a specific blockchain ecosystem. For example, in crypto currency, each computer must run on respective wallet applications to participate in its blockchain ecosystem.

**Common Ledger:** The ledger is a data structure that is managed by node application. When, the node applications running on the computer, the contents of the blockchain ecosystem become visible to the user. Further, there is only one common ledger for each ecosystem you participate in.

**Consensus Algorithm:** The consensus algorithm is a rule that defines how the ecosystem arrives at the distributed ledger. The various ecosystem has a different approach to reaching this consensus. There are mainly three methods that used to determine the participation in the consensus-building namely: Proof of work, Proof of stake, Proof of elapsed time.

**Virtual Machine:** The virtual machine is a computer program in which instructions are preloaded in the programming language. When an application running on the computer then these preloaded instructions make a change in the current state of the computer and these changes can be observed by different means for example like change in the information and graphics, sound on the speaker, change in the data, etc. provided by Albanese et al. (2020).

## Types of Blockchain

The blockchain is classified into three categories.

**Public Blockchain:** As the name indicates, public blockchain is an open network that can be accessible to all. In the public blockchain, data is visible to all and anyone can add new data. However, the public blockchain has the basic feature of a blockchain that is it is decentralized and immutable given by Yang et al. (2020).

**Private Blockchain:** Private blockchain is different from the public chain. The private blockchain is primarily used by an organization or agency and data is accessible to employees of that organization only. The organizations can set the rule for their private blockchain like who can access the data, who can add data, etc. provided by Du. Mingxiao et al. (2020).

**Consortium Blockchain:** Consortium blockchain or Federated blockchain is a platform controlled by a group of organizations or agencies. It is a blockchain in which different organizations shared data. The data is accessible to any organization that is part of a particular consortium blockchain. This blockchain allows better communication between the organization and enterprise given by Yang et al. (2020).

## Cloudmind and Implementation of Blockchain

The cloudmind or crowdmind is a virtual mind that is placed on the internet cloud. The cloudmind or crowdmind is the combination of different minds working together. There are different types of cloudmind namely:

**Machine Mind:** In the machine mind, the basic machine-based algorithm runs in the background based on Big data analysis.

**Machine + Human Mind:** In this cloudmind, both human and cloud-based machine thinking work together.

**Mindpool:** The mindpool is a combination of all the cloudminds such as machine-based mind, Human + Machine mind, Human mind.

The cloudmind is a cloud-based thinker with having kind of analytical power. However, there are many risks associated with cloudmind such as privacy leakage, insecurity, loss of personal identity, credit tracking.

## Blockchain Implementation

All the above mention drawbacks in the cloudmind can be overcome by using blockchain technology. The blockchain can act as an administrator for cloudmind. As the blockchain is secure and has a software structure that is ideal for cloudmind. Therefore, blockchain can serve as a trustworthy tool or technology that can be used to implement cloudmind. Further, blockchain has features of recording every transaction of an individual in a network which makes it ideal for cloudmind technology. The blockchain technology further implements anti-viral and anti- crowdmind in the cloudmind. There may be a feature that limits the amount of time an individual can spend on cloudmind. The blockchain can track the contribution of an individual in the cloudmind and therefore with the help of BCI technology if there is any prisoner-like behavior or neurodegeneration the system can generate a warning signal provided by Birbaumer et al. (2008), Swan et al. (2016), Statt et al. (2018).

# CONCLUSION

Brain-computer interfacing provides the path for communication with the external devices without the help of any muscular activity. The recent growth in this field attracted the research community. Several studies regarding the application of the BCI such as the medical, role of 5G technology in BCI, and blockchain implementation and authentication have been discussed in this paper. The further advancement in the BCI technology may allow one day to successfully implant brain chip in the neurons of the human brain and may allow the computerized simulate of the human brain that can be elongate at will. Further, Cloudmind or Crowdmind which is an essential application of BCI can be implemented with the help of blockchain. The cloudmind is a group of minds working together to solve a problem, to generate ideas, or for entertainment. Some challenges are associated with cloudmind. However, these challenges can be overcome with blockchain technology. The blockchain can work as a perfect tool to implement the cloudmind and can provide the necessary support.

## FUTURE SCOPE

The recent advancement shows that BCI is a dynamic and growing field. Since BCI is a versatile field. Therefore, the evolution of this field depends upon various factors such as new hardware technology, 5G technology, blockchain implementation, advancement in robotics and AI, machine learning, furtherance in medical and neuroscience, etc.

## REFERENCES

Abdulkader, S. N., Atia, A., & Mostafa, M.-S. M. (2015). Brain computer interfacing: Applications and challenges. *Egyptian Informatics Journal, 16*(2), 213–230.

Abdullah-Al-Mamun, K. (2013). *Pattern identification of movement related states in biosignals* (PhD diss.). University of Southampton.

Albanese, G., Calbimonte, J.-P., Schumacher, M., & Calvaresi, D. (2020). Dynamic consent management for clinical trials via private blockchain technology. *Journal of Ambient Intelligence and Humanized Computing, 11*(11), 1–18. doi:10.100712652-020-01761-1

Ang, S., Chee, M. S., Pepper, M., & Phillips, M. (2011). Use of brain computer interfaces in neurological rehabilitation. *British Journal of Neuroscience Nursing, 7*(3), 523–528. doi:10.12968/bjnn.2011.7.3.523

Birbaumer, N., Murguialday, A. R., & Cohen, L. (2008). Brain–computer interface in paralysis. *Current Opinion in Neurology, 21*(6), 634–638.

Birbaumer, Ruiz, & Sitaram. (2013). Learned regulation of brain metabolism. *Trends in Cognitive Sciences, 17*(6), 295–302.

Brown & Semelka. (2011). *MRI: Basic principles and applications*. John Wiley & Sons. https://www.cs.sfu.ca/~stella/papers/blairthesis/main/node11.html

BuiltIn. (n.d.). Retrieved from https://builtin.com/blockchain

Diamandis, P. H. (2016). *Nanorobots: Where we are today and why their future has amazing potential*. Singularity Hub. https://singularityhub.com/2016/05/16/nanorobots-where-we-are-today-and-why-their-future-has-amazing-potential/

Du, M., Chen, Q., Chen, J., & Ma, X. (2020). An optimized consortium blockchain for medical information sharing. *IEEE Transactions on Engineering Management.*

Fromherz, P. (2003). Neuroelectronic interfacing: Semiconductor chips with ion channels, nerve cells and brain. *Nanoelectronics and Information Technology.*

GPPP Architecture Working Group. (2017). *View on 5G Architecture: Version 2.0.* Author.

Gupta, S. S. (2017). *Blockchain.* Error! Hyperlink reference not valid.

Hai, A., Shappir, J., & Spira, E. (2010). Long-Term, Multisite, Parallel, In-Cell Recording and Stimulation by an Array of Extracellular Microelectrodes. *Journal ofNeurophysiology, 104*(1), 559–568. doi:10.1152/jn.00265.2010 PMID:20427620

Hazubski, S., Soekadar, S. R., Hoppe, H., & Otte, A. (2019). Neuroprosthetics 2.0. *EBioMedicine, 48,* 22. doi:10.1016/j.ebiom.2019.09.036 PMID:31594748

HistoryB. (n.d.). In *Wikipedia.* Retrieved from https://en.wikipedia.org/wiki/Blockchain#History

Horch, K. W., & Kipke, D. R. (Eds.). (2017). *Neuroprosthetics: Theory and practice* (Vol. 8). World Scientific. doi:10.1142/10368

Huettel, S. A., Song, A. W., & McCarthy, G. (2004). *Functional magnetic resonance imaging* (Vol. 1). Sinauer Associates.

Kim, T., Kim, S., Shin, D., & Shin, D. (2011). Design and implementation of smart driving system using context recognition system. In *2011 IEEE Symposium on Computers & Informatics* (pp. 84-89). IEEE. 10.1109/ISCI.2011.5958889

Lal, T. N., Schröder, M., Hill, N. J., Preissl, H., Hinterberger, T., Mellinger, J., & Bogdan, M. (2005). A brain computer interface with online feedback based on magnetoencephalography. In *Proceedings of the 22nd international conference on Machine learning* (pp. 465-472). ACM. 10.1145/1102351.1102410

Lin, C.-T., Lin, B.-S., Lin, F.-C., & Chang, C.-J. (2012). Brain computer interface-based smartliving environmental auto-adjustment control system in UPnP home networking. *IEEE Systems Journal, 8*(2), 363–370. doi:10.1109/JSYST.2012.2192756

Martins, Angelica, Chakravarthy, Svidinenko, Boehm, Opris, & Lebedev. (2019). Human brain/cloud interface. *Frontiers in Neuroscience, 13,* 112. https://www.frontiersin.org/articles/10.3389/fnins.2019.00112/full

Martone, R. (2019). *Scientists demonstrate direct brain-to-brain communication in humans*. Available online: https://www. scientificamerican. com/article/scientists-demonstrate-direct-brain-to-brain-communication-in-humans/

Mesgarani, N., & Chang, E. F. (2012). Selective cortical representation of attended speaker in multi-talker speech perception. *Nature*, *485*(7397), 233–236. doi:10.1038/nature11020 PMID:22522927

Nijboer, F., Furdea, A., Gunst, I., Mellinger, J., McFarland, D. J., Birbaumer, N., & Kübler, A. (2008). An auditory brain–computer interface (BCI). *Journal of Neuroscience Methods*, *167*(1), 43–50. doi:10.1016/j.jneumeth.2007.02.009 PMID:17399797

Nofer, M., Gomber, P., Hinz, O., & Schiereck, D. (2017). Blockchain. *Business & Information Systems Engineering*, *59*(3), 183–187. doi:10.100712599-017-0467-3

Obraczka, K., & Silva, F. (2000). Network latency metrics for server proximity. In *Globecom'00-IEEE. Global Telecommunications Conference. Conference Record* (Cat. No. 00CH37137) (Vol. 1, pp. 421-427). IEEE. 10.1109/GLOCOM.2000.892040

Potter, S. M., Wagenaar, D. A., & DeMarse, T. B. (2006). Closing the loop: stimulation feedback systems for embodied MEA cultures. In *Advances in network electrophysiology* (pp. 215–242). Springer. doi:10.1007/0-387-25858-2_9

PriceWaterhouseCoopers. (2016). *Making sense of bitcoin, cryptocurrency, and blockchain*. Author.

Ruiz, S., Buyukturkoglu, K., Rana, M., Birbaumer, N., & Sitaram, R. (2014). Real-time fMRI brain computer interfaces: Self-regulation of single brain regions to networks. *Biological Psychology*, *95*, 4–20. doi:10.1016/j.biopsycho.2013.04.010 PMID:23643926

Rutten, W. L. (2002). Selective Electrical Interfaces with the Nervous System. *Annual Review of Biomedical Engineering*, *4*(1), 407–452. doi:10.1146/annurev.bioeng.4.020702.153427 PMID:12117764

Shien, Y. K. (2019). *5G: Enabling technologies & applications*. Academic Press.

Statt, N. (2018). *Kernel is trying to hack the human brain–but neuroscience has a long way to go*. The Verge. Available at: www. theverge. com/2017/2/22/14631122/kernel-neuroscience-bryan-johnsonhuman-intelligence-ai-startup

Study: Monkey brains move cursor. (2002). *CNN*. Retrieved from https://www.ai.rug.nl/~lambert/projects/BCI/literature/misc/monkey-3d-ball-thinker.html

Švogor, I., & Kišasondi, T. (2012). Two factor authentication using EEG augmented passwords. In *Proceedings of the ITI 2012 34th International Conference on Information Technology Interfaces* (pp. 373-378). IEEE.

Swan, M. (2016). The future of brain-computer interfaces: Blockchaining your way into a cloudmind. *Journal of Ethics and Emerging Technologies, 26*(2), 60–81.

Tan, D., & Nijholt, A. (2010). Brain-computer interfaces and human-computer interaction. In *BrainComputer Interfaces* (pp. 3–19). Springer. doi:10.1007/978-1-84996-272-8_1

Vecchiato, G., Babiloni, F., Astolfi, L., Toppi, J., Cherubino, P., Dai, J., Kong, W., & Wei, D. (2011). Enhance of theta EEG spectral activity related to the memorization of commercial advertisings in Chinese and Italian subjects. In *2011 4th International Conference on Biomedical Engineering and Informatics (BMEI)* (Vol. 3, pp. 1491-1494). IEEE. 10.1109/BMEI.2011.6098615

Waldert, S. (2016). Invasive vs. non-invasive neuronal signals for brain-machine interfaces: Will one prevail? *Frontiers in Neuroscience, 10*, 295. doi:10.3389/fnins.2016.00295 PMID:27445666

Wei, L., Hong, Q., Yue, H., & Xi, C. (2010). The research in a plantar pressure measuring system connected with EEG. In *IEEE 10th International Conference on Signal Processing Proceedings* (pp. 434-437). IEEE. 10.1109/ICOSP.2010.5655374

Yang, R., Wakefield, R., Lyu, S., Jayasuriya, S., Han, F., Yi, X., Yang, X., Amarasinghe, G., & Chen, S. (2020). Public and private blockchain in construction business process and information integration. *Automation in Construction, 118*, 103276. doi:10.1016/j.autcon.2020.103276

Zhang, X., Yao, L., Wang, X., Monaghan, J., Mcalpine, D., & Zhang, Y. (2019). *A survey on deep learning based brain computer interface: Recent advances and new frontiers*. arXiv preprint arXiv:1905.04149

Ziegler-Graham, K., MacKenzie, E. J., Ephraim, P. L., Travison, T. G., & Brookmeyer, R. (2008). Estimating the prevalence of limb loss in the United States: 2005 to 2050. *Archives of Physical Medicine and Rehabilitation, 89*(3), 422–429. doi:10.1016/j.apmr.2007.11.005 PMID:18295618

# Chapter 5
# Role of Graph Theory in Computational Neuroscience

**Hitesh Marwaha**
*GNA University, India*

**Anurag Sharma**
https://orcid.org/0000-0003-0926-1699
*GNA University, India*

**Vikrant Sharma**
*GNA University, India*

## ABSTRACT

*Neuroscience is the study of the brain and its impact on behavior and cognitive functions. Computational neuroscience is the subfield that deals with the study of the ability of the brain to think and compute. It also analyzes various electrical and chemical signals that take place in the brain to represent and process the information. In this chapter, a special focus will be given on the processing of signals by the brain to solve the problems. In the second section of the chapter, the role of graph theory is discussed to analyze the pattern of neurons. Graph-based analysis reveals meaningful information about the topological architecture of human brain networks. The graph-based analysis also discloses the networks in which most nodes are not neighbors of each other but can be reached from every other node by a small number of steps. In the end, it is concluded that by using the various operations of graph theory, the vertex centrality, betweenness, etc. can be computed to identify the dominant neurons for solving different types of computational problems.*

DOI: 10.4018/978-1-7998-7433-1.ch005

## INTRODUCTION TO NEUROSCIENCE

The only discipline that can help you understand how you think and process knowledge in your brain is computational neuroscience. Computational neuroscience's ultimate aim is to understand how electrical and chemical signals in the brain are used to represent and process information. It covers biophysical computation processes in neurons, computer simulations of neural circuits, and learning models. In general, Before the start of any routine task say reading a book, going to the office or going to market, etc. there are few perceptions that our brain might consider, like why to read or go, What would be the outcome, etc. It's all the mathematics, permutations, and chemical equations that are going on within our brains. Eric L. Schwartz coined the term "computational neuroscience" at a conference to provide a study of a discipline that had previously been known by several names, including "neural modeling," "brain theory," and "neural networks." Neuroscience covers a wide range of disciplines, including molecular and cellular research as well as human psychophysics and psychology. Computational neuroscience aims to explain how the brain interprets and processes information using electrical and chemical signals. Although this goal is not recent, a lot has changed in the last decade. Because of developments in neuroscience, more information about the brain is now available, more computational power is available for practical simulations of neural systems, and new perspectives are being derived from the analysis of simplified models of large networks of neurons. The challenge of understanding the brain is attracting a rising number of scientists from various disciplines. Although there has been an explosion of findings of the structure of the brain at the cellular and molecular levels over the last few decades, we still don't know how the nervous system allows us to see, hear, learn, remember, and plan those acts. Computational neuroscience is a branch of neuroscience that creates models to incorporate large amounts of data in order to better understand how the brain works(Teeters et al., 2008).

A growing number of neuroscience databases are being created and made publicly accessible on the internet. In particular, there has been important progress in the collection and representation of neuroanatomical association patterns. Researchers can now access extensive data sets of connectional relationships between individual neurons or brain areas thanks to these efforts. The availability of such data sets necessitates the development of appropriate analytical resources for thorough and principled research. Graph theory, a branch of mathematics and combinatorics with many applications in various fields, such as R. Kötter (ed.), Neuroscience Data Analysis, provides one avenue for such an analysis(Sporns, 2003).

While using similar analytical methods, the computational neuroscience and systems biology research groups have relatively little contact. In this study, I reconstruct the past of the two disciplines and argue that this helps to understand why they grew up so far apart. It's a shame that they're separated because both fields would benefit from each other. Several examples are given, ranging from sociological to software technological to methodological. Computational neuroscience has more experience with multiscale modeling and the study of information processing by biological systems, whereas systems biology, is a better-structured culture that is very good at sharing resources(De Schutter, 2008).

## GRAPH THEORY

Networks and graphs are sets of elements (nodes, vertices) and their pairwise links (edges, connections) that can be summarised in the form of a relation (or adjacency) matrix in their most basic form. The topology of a graph is defined by the complete set of all pairwise connections, which provides a complete map of all node and edge relationships. Depending on the measuring tool, brain nodes may be individual neurons or whole brain regions. Edges can be binary or weighted, and they can be guided or undirected, depending on how empirical evidence is used to estimate interactions. The essence of the edge representation must be considered when choosing suitable graph theory methods for modeling and analyzing empirical data. To put it another way, not all graph theory approaches are appropriate for all applications(Battiston et al., 2017). Graph theory is commonly used in biological mathematics to solve a variety of biology problems as an important modeling, analysis, and computational method. In the field of microbiology, a graph may be used to represent the molecular structure, with a vertex representing a cell, gene, or protein, and an edge representing the linked part(Gao et al., 2018).

The area of mathematics is extremely significant in a variety of fields. Graph theory, which is used in structural models, is an important field of mathematics. This structural planning of different objects or innovations leads to new developments and changes to the current environment in order to advance in those fields. In 1735, the advent of the Konigsberg bridge problem launched the field graph theory (Vedavathi & Gurram, 2013). Computer science applications make extensive use of graph theory concepts. Especially in computer science research areas such as data mining, image segmentation, clustering, image capturing, networking, data science, cyber, studying the human brain anatomical network via graphs (Iturria-medina et al., 2008), pattern recognition (Siqueira et al., 2014), and so on, For instance, a data structure could be constructed in the form of a tree, with vertices and edges. Graph principles can be used to model network topologies as well. The most important principle of graph coloring is often used in resource allocation and setup. In addition, graph theory's paths walk, and circuits are used in a variety of applications, including the traveling salesman problem, database design principles, and resource networking. Techniques of graph theory are commonly used in a variety of fields, including chemistry, biology, physics, and computer science. GT is a mathematical notation with widespread modeling and research applications in every discipline (Farahani et al., 2019).

A graph is an ordered pair of points called nodes and the lines joining the nodes, called edges of the graph i.e. of the form (node, edge). The edges are lines or arcs that connect any two/multiple nodes in a graph, and the vertices/vertex are the nodes. For example, is G(U, V), where U is the list of users and V is the set of edges indicating the relationship between the users or objects. There are two types of graphs: directed and undirected. The edges of undirected graphs have no direction. Figure 1 shows an example of an undirected graph.

*Figure 1.*

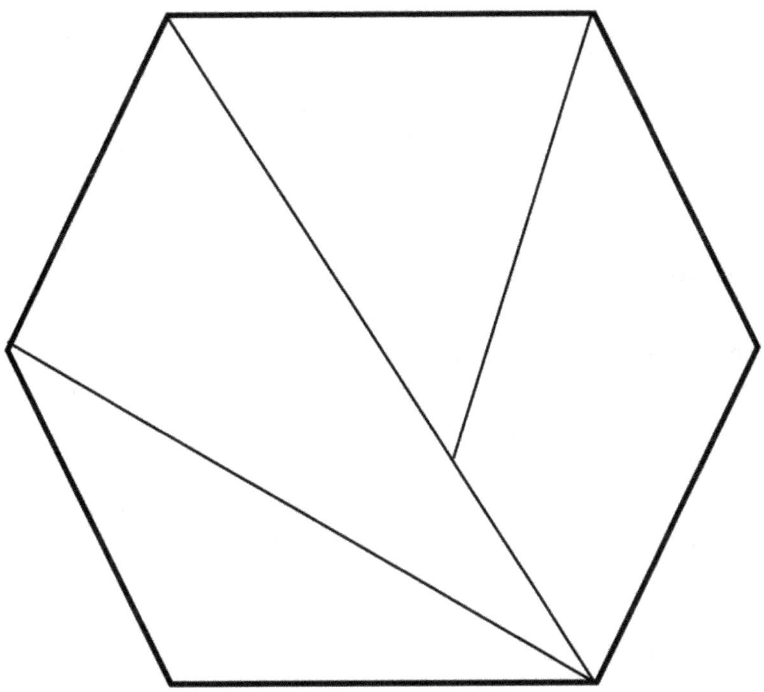

People and friendship in a social network, or researchers and co-authored papers in a collaboration network, are probable examples of the undirected graph. In directed graphs, the edges have a specific path (i.e., they can be considered incoming or outgoing). In these cases, the graphs are drawn with arrowheads on the edges. Digraphs are the most common name for directed graphs. Figure II shows an example of a directed graph.

*Figure 2.*

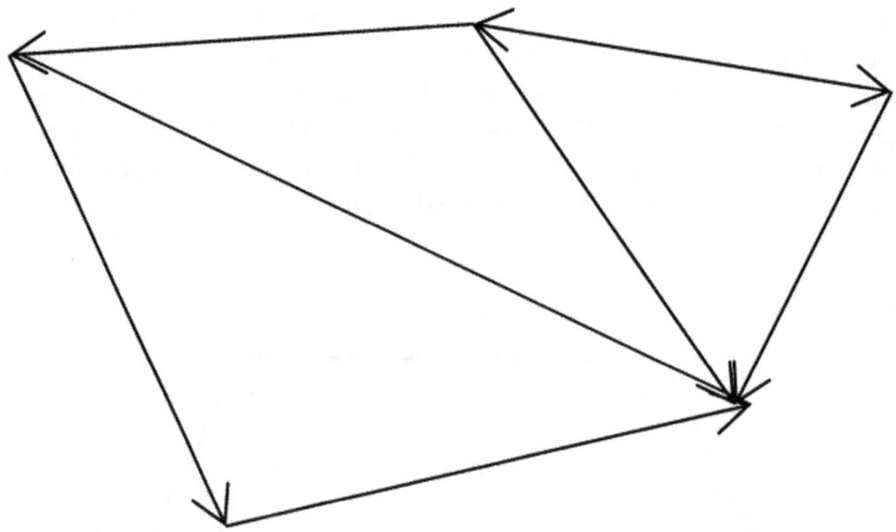

## Data Structures and Representations

An adjacency matrix can be used to represent a graph in the memory of the computer. An adjacency matrix A is a square matrix of size N X N (where N is the number of vertices) that is used to describe a graph in graph theory. The adjacency matrix for a simple graph is a (Sabidussi, 1966; Yue et al., 2019)-matrix with zeros on its diagonal (A[i,j] = 1 for connection presence, A[i,j] = 0 for connection absence) or a (0,wij)-matrix for a weighted graph (A[i,j] = wij).

## WHY GRAPH FOR NEUROSCIENCE

It is now possible to investigate the structure-function relationships in the brain due to the development of neuroimaging techniques. When the brain is not engaged in any cognitive activity or activated by any external input, it retains essential tasks that are distributed in well-defined spatial patterns. It has recently been proposed to model the observed functional pattern from the arrangement of white matter fiber bundles in order to understand the self-organization of the brain from its anatomical structure.

Graph theory is often used in neuroscience to investigate how various areas of the brain (nodes) are functionally related. We'll concentrate on modeling functional brain connectivity with undirected graphs (also known as "static" graphs in neuroscience. Functional magnetic resonance imaging is one way to test functional brain activity (fMRI). The blood pressures in various areas of the brain are measured using this imaging method. It uses the magnetic susceptibilities of oxyhemoglobin and deoxyhemoglobin, two molecules in the blood, to determine their amounts. The amount of blood in a brain area determines whether it is activated or "functional." This is because capillaries supply more energy to neurons as they activate (i.e. the part of the brain is functioning), a mechanism known as the "hemodynamic response."

When two areas of the brain are "functionally related," it suggests that the measurements of action recorded by them have a meaningful connection. That is, as one brain region's activity rises, the activity of the other rises as well. This type of connectivity between two areas can be expressed as the node and edge relationships for finding some useful analysis for the future. We normally calculate the Pearson correlation coefficient r between the two brain signals (Medium, n.d.).

$$r = \frac{\sum (x-\bar{x})(y-\bar{y})}{\sqrt{\sum (x-\bar{x})^2 \sum (y-\bar{y})^2}}$$

Here graph theory comes into existence again. To represent functional components of the brain as a node of the graph and each link between these nodes(edge) is represented by the calculated value of r.

As already know the value of r lies between -1 and 1. In figure III circles represent nodes and lines represent edges.

*Figure 3.*

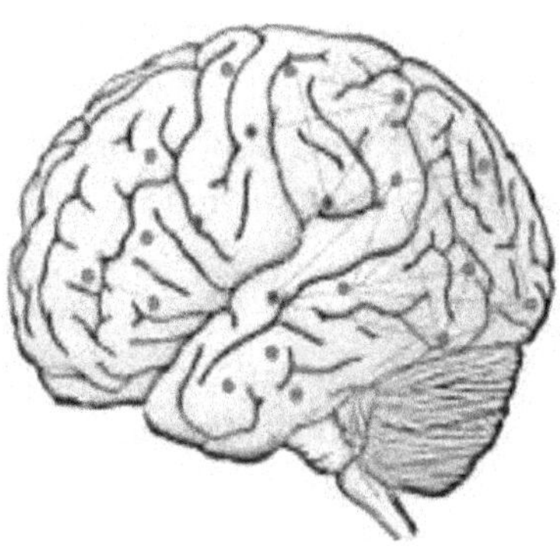

Now the aforementioned formula of correlation the value of r is used to draw the graph. In this case, we do not consider the value less than 0. The zero value itself depicts the absence of an edge between two components of the brain. To represent the neurons and signals using the graph, if there is any functional connectivity between two neurons there will be some value of r greater than zero less than or equal to one. The value one represents the strong functional connectivity between the neurons and in the graph that can be represented as a thick edge.

## Clustering Coefficient in Graph

A clustering coefficient is a measure of how closely nodes in a graph cluster together. Evidence shows that in most real-world networks, especially social networks, nodes appear to form highly cohesive networks with a high density of connections; this possibility is higher than the average chance of a connection forming arbitrarily between two nodes. Node degree, sometimes abbreviated as is one of the most basic and essential metrics for a brain network. A node's degree is the number of edges that link it to all other nodes. In general, the higher a node's degree, the more nodes it is attached to and the more important it is in the brain network (Liu et al., 2017). In a binary undirected network, the degree $C_i$ of a node is defined as (Rubinov & Sporns, 2010)

$$C_i = \sum A_{ij}$$

where if $A_{ij}$ if the connection of node i and node j exists; otherwise where if the connection of node and node exists; otherwise $A_{ij} = 0$

## DIFFERENT TYPE OF ANALYSIS DONE USING GRAPH

- Centrality analysis: A node or object is central if it's in the middle of a larger entity. In real life, something is central if it's important and vital. For graphs, identify important nodes by looking at how central they are. In the graph. In the graph, the green node is the central node.

Some nodes in a graph play a critical role and are referred to as "central" within the network structure. Central nodes can be defined using centrality metrics, which include "degree, betweenness, and eigenvector centrality, among others". The most connected nodes are identified by degree, while those located on the most followed paths are identified by betweenness centrality (Joyce, 2010).

*Figure 4.*

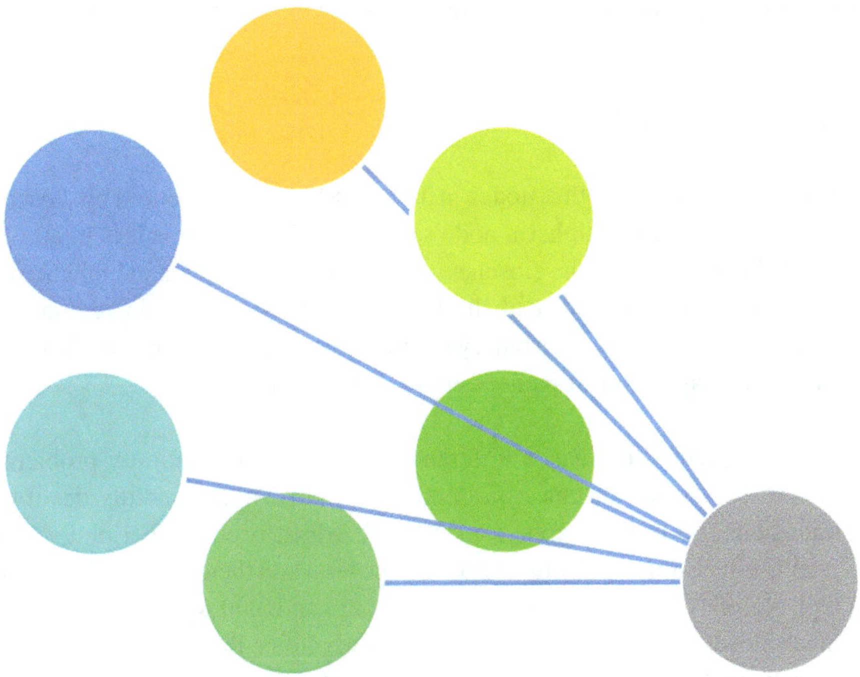

The centrality of nodes in a neural network is used to determine their significance. The greater a node's centrality, the more powerful it is in the brain network's information transmission. Three measurements of centrality are often used in brain network research, including degree centrality, closeness centrality, and betweenness centrality, as follows:

**Degree Centrality in Brain Network:** The central degree is the most popular centrality measure which uses the node degree to define the node's importance in a brain network. The centrality level of a brain region tests the direct influence of the brain region on other neighboring brain regions during brain network research. As already mentioned node degree defines the centrality of a node so when we use graphical representation first to calculate $\sum$the degree of each node. The degree is defined as the number of edges incident to a particular node in the graph. So the degree of each node can be calculated by the following formula:

Let $A_{ij}$ is an adjacency matrix of the given graph then the degree of each node can be calculated as ;

$$D(i) = \sum A_{ij} \ (i \neq j) \ i = 1 \text{ to } n$$

$$C(n) = \text{Max } (D(i))$$

Hence C(n) represents the central node in the graph. Hence the graphical representation of neural structure can be used to find the central node in the brain network.

## Closeness Centrality

The closeness of a node to other nodes in a brain network is measured by closeness centrality. As a result, the higher a node's centrality is, the closer it is to all other nodes in the brain network. The closeness centrality of a brain area in brain network analysis tests the indirect effect of the brain region on other brain regions. The closeness centrality of a node in a binary undirected network is defined as the inverse of the node's average shortest path length to all other nodes.

- Closeness centrality takes a different approach to the centrality problem. It acts as the shortest distance of a node to all other nodes and divides it by a minus one. So in the graph, a node like I, which is on the periphery of the graph, will be quite far from all nodes in general and therefore will have a very high closeness centrality value. On the other hand, nodes like F, C, and H are much closer to all other nodes. Know that we define this measure in terms of shortest paths. So if a moving object, like information, flows to the shortest path in the network, F is more likely to receive them earlier than other nodes. And therefore can process into other nodes more quickly. Therefore. If we look to inject a new piece of information into the network with the idea that it should read every other node quickly, I should possibly inject it at F. Recall, however, that node information flow is two shortest routes. An example is a gossip, which tends to travel through centrality nodes.

Closeness centrality does not work well for these types of information nodes. Another very popular centrality measure is called betweenness centrality. For any node, it measures the fraction of shortest paths flowing through that node. Compared to the number of shortest paths in the graph. Since B is at one end, it's between a centrality is 0. But let's look at A. A is in the path from B to E. So is D. Therefore, A's score for the B to E path is 0.5. Similarly, its score for the C to E path is also 0.5, making the total score 1. E is in the path from A to D, for there is one more path from A to do through C, so E's score is 0.5. Now can you verify why C's score is 3.5? Betweenness centrality is typically used for the problem where a commodity is flowing through the network. As in the case of closeness centrality. Any quantity that does not flow in shortest path channels like infection or rumor on the internet doesn't work well with betweenness centrality.

*Figure 5.*

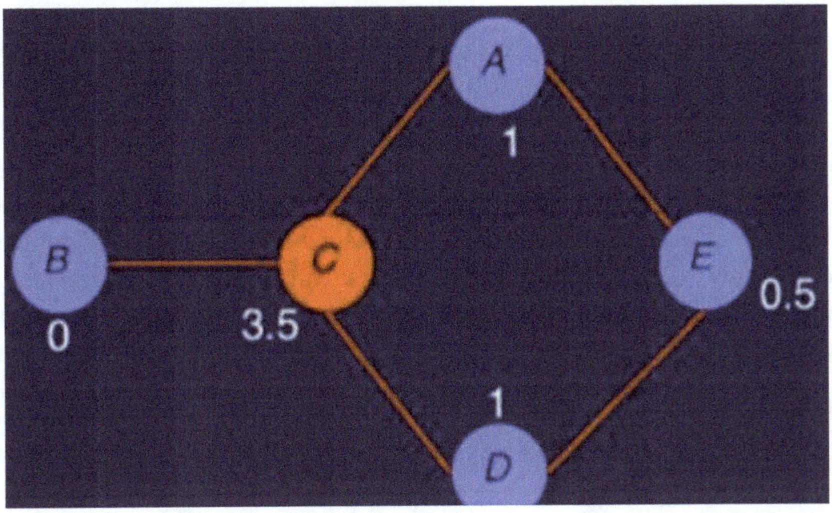


## GRAPH GENERATOR DATABASES

As discussed in previous sections graph analytics plays important role in computational neuroscience. In neuroscience, graph theory is frequently used to examine how different parts of the brain (nodes) are functionally connected. To combine many different data sources and increase the understandability and usability of neuroscience data, a graph-based approach for representing, analyzing, and accessing brain-related data (Angles & Gutiérrez, 2008) might be utilized. To model the data and analyze the same for various analytics like centrality, the closeness of nodes, and betweenness some kind of software tool is required. Graph data models and associated graph database management systems provide performance, flexibility, and agility, as well as the ability to use well-established graph analytics solutions; however, there is limited research on graph-based data representation as a mechanism for the integration, analysis, and reuse of neuroscience data(Ercim, n.d.). Several data modeling tools are available these days. RDBMS helps in creating the logical model that creates the tabular structure of data and then helps in easy retrievals of information. A graph model may be used to design and implement computational neuroscience data, and the same data may be transferred from a relational database to a NoSQL graph database (Cattell, 2011). However, modeling the data graph is not very simple as traditional database systems (Neo4j, n.d.).Neo4j is an ideal production-ready graph database tool that can efficiently model queries about complex and highly connected graph data (Vukotic, 2014).

## CONCLUSION

The exponential rise of network neuroscience over the last two decades has been nothing short of miraculous. The availability of relational data reflecting couplings and interactions among parts of brain systems is a primary driving force behind this rapid expansion. For the time being, most research is descriptive and focuses solely on paired interactions that result in graphs with relational linkages. Graph theory, on the other hand, is far more powerful than existing techniques for analyzing brain networks imply. General models, dynamic networks, multilayered models, and algebraic topology are just a few of the potential approaches being studied at the moment. These new methodologies are expected to find use not just in basic research, but also in scientific and clinical research in the future. Graph theory approaches will continue to be essential tools in improving our understanding of the brain as a complex interrelated system for many years to come.

## REFERENCES

Angles, R., & Gutiérrez, C. (2008). Survey of graph database models. *ACM Computing Surveys*, *40*(1), 1–39. doi:10.1145/1322432.1322433

Battiston, F., Nicosia, V., Chavez, M., & Latora, V. (2017). Multilayer motif analysis of brain networks. *Chaos*, *27*(4), 047404. doi:10.1063/1.4979282 PMID:28456158

Cattell, R. (2011). Scalable SQL and NoSQL data stores". *SIGMOD Record*, *39*(4), 12–27. doi:10.1145/1978915.1978919

De Schutter, E. (2008). Why are computational neuroscience and systems biology so separate? *PLoS Computational Biology*, *4*(5), e1000078. doi:10.1371/journal.pcbi.1000078 PMID:18516226

Ercim. (n.d.). https://ercim-news.ercim.eu/en125/r-i/graph-based-management-of-neuroscience-data-representation-integration-and-analysis

Farahani, F. V., Karwowski, W., & Lighthall, N. R. (2019). Application of Graph Theory for Identifying Connectivity Patterns in Human Brain Networks: A Systematic Review. *Frontiers in Neuroscience*, *13*, 585. doi:10.3389/fnins.2019.00585 PMID:31249501

Gao, W., Wu, H., Siddiqui, M. K., & Baig, A. Q. (2018). Study of biological networks using graph theory. *Saudi Journal of Biological Sciences*, *25*(6), 1212–1219. doi:10.1016/j.sjbs.2017.11.022 PMID:30174525

Iturria-medina, Y., Sotero, R. C., Canales-rodríguez, E. J., Alemán-gómez, Y., & Melie-garcía, L. (2008). Studying the human brain anatomical network via diffusion-weighted MRI and Graph Theory. *NeuroImage, 40*(3), 1064–1076. doi:10.1016/j.neuroimage.2007.10.060 PMID:18272400

Joyce, K. E. (2010). A new measure of centrality for brain networks. *PloS One, 5*(8).

Liu, J., Li, M., Pan, Y., & Lan, W. (2017). Complex Brain Network Analysis and Its Applications to Brain Disorders: A Survey. Complexity. doi:10.1155/2017/8362741

Medium. (n.d.). https://medium.com/swlh/graph-theory-machine-learning-in-neuroscience-30f9bec5d182#:~:text=In%20neuroscience%2C%20we%20often%20use,functionally%20connected%20to%20each%20other.&text=We'll%20be%20focusing%20on%20resting%2Dstate%20functional%20connectivity%2C,completing%20any%20kind%20of%20task

Neo4j. (n.d.). https://neo4j.com/blog/data-modeling-basics/

Rubinov, M., & Sporns, O. (2010). Complex network measures of brain connectivity: Uses and interpretations. *NeuroImage, 52*(3), 1059–1069. doi:10.1016/j.neuroimage.2009.10.003 PMID:19819337

Siqueira, S., Eduardo, C., Junior, B., Comfort, W. E., Rohde, L. A., & Sato, J. R. (2014). Abnormal Functional Resting-State Networks in ADHD: Graph Theory and Pattern Recognition Analysis of fMRI Data. *BioMed Research International, 2014*, 380531. PMID:25309910

Sporns, O. (2003). Graph theory methods for the analysis of neural connectivity patterns. In *Neuroscience databases* (pp. 171–185). Springer. doi:10.1007/978-1-4615-1079-6_12

Teeters, J. L., Harris, K. D., Millman, K. J., Olshausen, B. A., & Sommer, F. T. (2008). Data sharing for computational neuroscience. *Neuroinformatics, 6*(1), 47–55. doi:10.100712021-008-9009-y PMID:18259695

Vedavathi & Gurram. (2013). Applications on Graph Theory. *International Journal of Engineering Research & Technology, 2*(1).

Vukotic, A. (2014). *Neo4j in action*. Manning Publications Co.

# Chapter 6
# Tractography of White Matter Fibers Using Diffusion MRI

**Strivathsav Ashwin Ramamoorthy**
https://orcid.org/0000-0003-2195-3428
*National Institute of Technology, Trichy, India*

## ABSTRACT

*To understand more about the human brain and how it works, it is vital to understand how the neural circuits connect different regions of the brain. The human brain is filled predominantly with water and the majority of the water molecules undergo diffusion which can be captured with the help of diffusion MRI. Diffusion weighted images enable us to reconstruct the neural circuits in a non-invasive manner, and this procedure is referred to as tractography. Tractography aids neurosurgeons to understand the neural connectivity of the patient. This chapter attempts to explain the procedure of tractography and different types of algorithms.*

## INTRODUCTION

The brain is one of the complex organs present in a human being and more research is being done to unravel the mysteries of the human brain. A lot of research has been into studying the structure of the brain and its function. In the past decades, we had the MRI (Magnetic Resonance Imaging) manipulated to understand the anatomy, the functional areas of the brain, and how the neurons inside the brain are connected. MRI has offered many valuable inputs to the doctor if they are to operate on a patient. For instance, doctors use a form of MRI called fMRI (functional magnetic resonance imaging) to observe, which areas of the brain receive oxygen-rich blood and try to identify the function associated with that region. Other than

DOI: 10.4018/978-1-7998-7433-1.ch006

fMRI many different types of MRI are sMRI and dMRI. sMRI (Structural MRI) helps to understand the anatomy of the human brain and dMRI (diffusion MRI) captures the diffusion of the water molecules. Here, in this chapter let us see how we can use dMRI, to reconstruct the neurons present in the brain.

The main principle of dMRI is restricted diffusion of water molecules takes place due to the presence of axon bundles in neurons. Based on the diffusion behavior captured using MRI, the neurons are reconstructed in a non-invasive manner. This process of reconstructing the neurons based on diffusion data is referred to as tractography. Apart from white matter fibers, tractography has been used to construct the muscle fibers.

dMRI has many overarching applications. Doctors can use it to infer the neural connectivity of the brain, before operating on a patient diagnosed with a tumor. In that case, the doctor has to dissect the tumor from the brain, and before that understanding, the neural structure is vital as they should not end up damaging critical neurons while doing dissection. A lot of acquisition schemes and tractography algorithms have been developed to improve the accuracy of reconstructing white matter fibers. In this chapter, we will discuss the tractography algorithms which have been developed over time.

## BACKGROUND

It all started, when Einstein in 1905 did an investigation on the random motion of water molecules. The human brain is filled predominantly with water and also has other fluids such as cerebrospinal fluid. Random motion of water molecules is observed which is referred to as Brownian motion. Fast forward big MRI machines with the help of magnetic fields created dMRI images. In the case of dMRI images, the MRI signal containing diffusion information is sensitized to the direction in which the magnetic field is being applied.

The arrangement of the white matter fibers in our brain gives rise to the diffusion signal and the acquired image is referred to as diffusion-weighted image (DWI). Scientists developed gradient schemes when entered on the MRI machine captures the diffusion behavior of water molecules. There are schemes such as one-shell acquisition, where only one magnetic field is applied, whereas in multi-shell acquisitions more than one magnetic field is used to acquire the dMRI image. For instance, if the dMRI image is acquired using only one magnetic field b = 1000 s/mm$^2$ then it is referred to as one shell dataset and if the dMRI image is acquired using more than one magnetic field such as b = 1000 s/mm$^2$ and b = 3000 s/mm$^2$ then the dataset is referred to as multi-shell acquisition. In the MRI sequence developed for capturing the dMRI image, a large number of gradient directions are defined. More the number of gradient directions more accurately the tractography algorithms can reconstruct the white matter fibers.

In clinical settings, one of the majorly used methods for the reconstruction of white matter fibers is Diffusion Tensor Imaging. In DTI, the tracking of white matter fibers is done by calculating the eigenvectors in each voxel of the brain. Voxel is analogous to what pixel is in a 2-Dimensional image. DTI method has its limitations, which we will see in detail in this chapter and understand why the other methods perform better than DTI.

## DATA ACQUISITION

Diffusion-weighted image is acquired with the help of MRI sequences written specifically to capture the diffusion behavior of the water molecules. The magnetic gradient is applied in a different direction as more the number of directions, more will be the accuracy in the reconstruction of white matter fibers. dMRI dataset which has more than 64 directions is referred to as HARDI (High Angular Resolution Diffusion Imaging) dataset. The standard dMRI dataset consists of many 3D volumes in one .nii file, where the number of 3D volumes is equal to the different directions in which the magnetic field is applied. For each gradient direction, we obtain a unique diffusion-weighted 3D image. Apart from the .nii file, we have two files called *bvecs* and *bvals*, where *bvecs* contains the X, Y and Z coordinates towards which the magnetic field is directed and *bvals* contains the magnitude of the magnetic field applied. For instance, the data used for the ISMRM tractography challenges had 32 different directions in which the magnetic field was applied and the magnitude of the magnetic field is $b = 1000\ s/mm^2$. For 32 different directions, the same number of unique 3D diffusion-weighted images was obtained. The below table captures the information of the magnitude of the magnetic field and the vector coordinates of X, Y, and Z of the magnetic field related to the ISMRM 2015 dataset.

*Table 1. Value of magnetic field, bvals and bvecs for the volumes of dMRI*

| Volume No: | b value | bx | by | bz |
|---|---|---|---|---|
| 1 | 1000 | -0.13 | -0.48 | 0.86 |
| 2 | 1000 | -0.87 | 0.26 | 0.408 |
| 3 | 1000 | -0.59 | -0.76 | -0.22 |
| 4 | 1000 | -0.25 | -0.68 | -0.68 |
| 5 | 1000 | -0.26 | 0.0050 | 0.96 |
| 6 | 1000 | -0.88 | 0.47 | 0.009 |
| 7 | 1000 | -0.85 | -0.25 | 0.44 |
| 8 | 1000 | -0.28 | 0.57 | -0.76 |
| 9 | 1000 | -0.39 | -0.90 | 0.14 |
| 10 | 1000 | -0.80 | -0.57 | 0.13 |
| 11 | 1000 | -0.45 | 0.44 | 0.77 |
| 12 | 1000 | -0.16 | 0.13 | -0.97 |
| 13 | 1000 | -0.177 | 0.75 | 0.62 |
| 14 | 1000 | -0.63 | 0.65 | 0.40 |
| 15 | 1000 | -0.18 | -0.93 | -0.30 |
| 16 | 1000 | -0.78 | -0.102 | -0.61 |
| 17 | 1000 | -0.67 | 0.07 | 0.73 |
| 18 | 1000 | -0.41 | -0.23 | -0.88 |
| 19 | 1000 | -0.59 | 0.80 | -0.05 |
| 20 | 1000 | -0.63 | -0.51 | -0.57 |
| 21 | 1000 | -0.54 | -0.67 | 0.50 |
| 22 | 1000 | -0.28 | 0.92 | 0.23 |
| 23 | 1000 | -0.11 | -0.82 | 0.55 |
| 24 | 1000 | -0.93 | 0.17 | -0.32 |
| 25 | 1000 | -0.03 | 0.99 | -0.12 |
| 26 | 1000 | -0.01 | 0.37 | 0.92 |
| 27 | 1000 | -0.91 | -0.34 | -0.22 |
| 28 | 1000 | -0.70 | 0.54 | -0.45 |
| 29 | 1000 | -0.57 | 0.24 | -0.77 |
| 30 | 1000 | -0.33 | 0.84 | -0.41 |
| 31 | 1000 | -0.99 | -0.03 | 0.08 |
| 32 | 1000 | -0.53 | -0.34 | 0.77 |

## WHITE MATTER FIBER BUNDLES

Collection of neurons are referred to as white fiber bundles. These fiber bundles are responsible for connecting different regions of the brain. For instance, the fiber bundle Corpus Callosum connects the left and right hemispheres of the brain.

*Figure 1. Corpus Callosum fiber bundle*

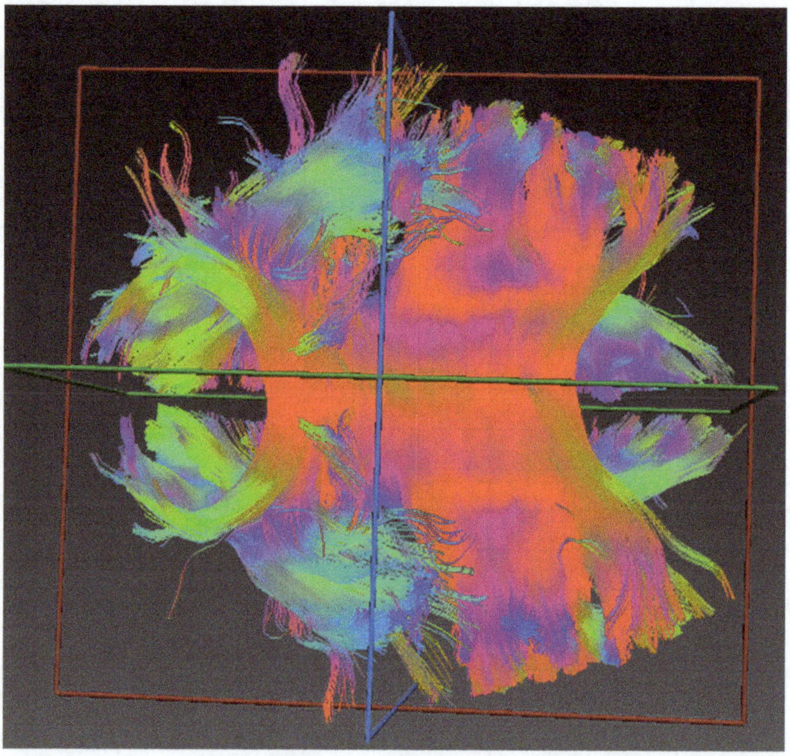

According to ISMRM 2015 tractography challenge around 25 white matter fiber bundles have been segmented manually. When all the white matter fibers are combined, we get the whole connectivity of how the white matter fibers are arranged in the brain.

*Figure 2. Coronal view*

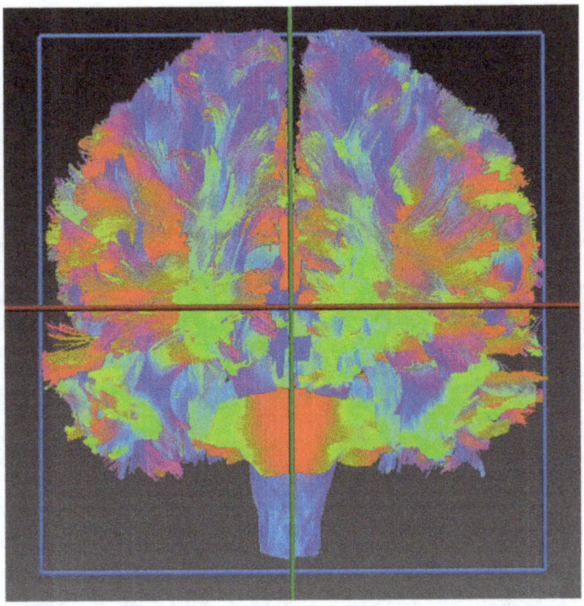

*Figure 3. Sagittal view*

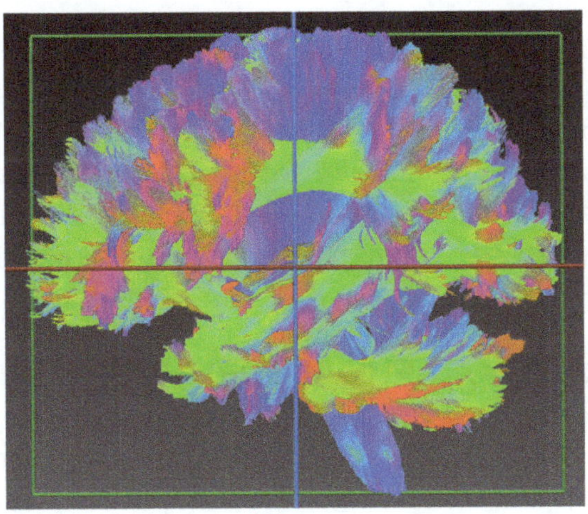

*Figure 4. Axial view*

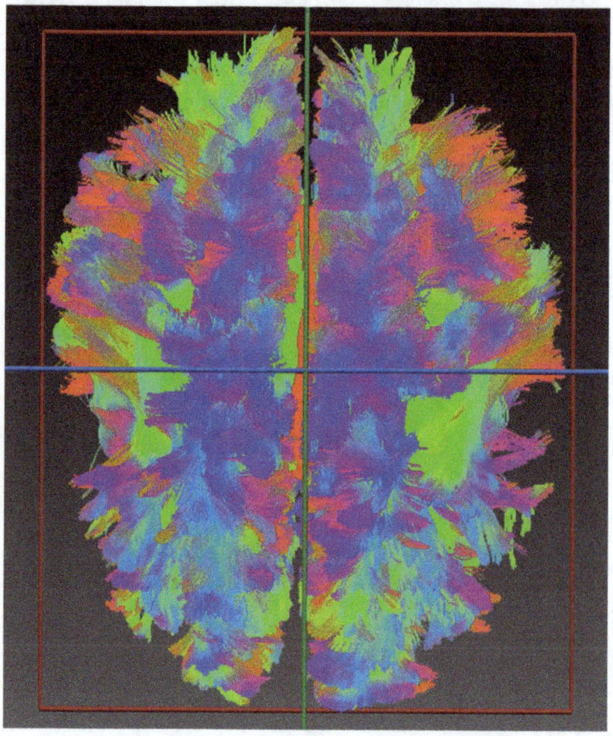

In the above figures, you can see the fibers being represented using different colors such as red, green, and blue from different views such as axial, sagittal and coronal. These colors symbolize the principal direction in which diffusion takes place. These kinds of images are also referred to as color-coded direction maps. If the color is red then the diffusion is along the sagittal axis, if it is blue then the diffusion is along the axial axis and if it is green then the diffusion is along the coronal axis.

## TYPES OF TRACTOGRAPHY ALGORITHMS

According to (Itay & Tammy, 2019), current tractography algorithms can be divided into deterministic, probabilistic, and global algorithms. A deterministic algorithm like DTI (Diffusion Tensor Imaging) tracks the fibers based on the direction of the principal vector. Apart from DTI, we have algorithms such as Q-ball imaging and spherical deconvolution. These models are based on some specific assumptions and therefore impose specific requirements on the data and acquisition protocol.

Recently, a lot of machine learning and deep learning techniques have been used for tractography. Since these algorithms are data-driven there is no specific requirement imposed on the data acquisition. (Peter F. Neher.et.al, 2015) used random forest classifier to track the white matter fibers. The classifier predicted a local fiber orientation based on the neighboring dMRI values of a voxel. (Philippe Poulin et al., 2017) made use of deep learning for tractography applications. Streamline tractography was addressed as a regression model by predicting the principal vector directions based on the sequence of dMRI values. (Itay & Tammy, 2019) proposed a deep learning framework to predict fODF (Fiber orientation distribution function) and do streamline tractography using the data from ISMRM 2015 tractography challenge. They performed a sequential classification task with the help of an RNN model to predict local fiber orientations along the streamlines.

## DETERMINISTIC TRACTOGRAPHY ALGORITHM - DIFFUSION TENSOR IMAGING

DTI which stands for Diffusion tensor imaging is a method of reconstructing neurons using DWI (Diffusion-weighted image). DTI provides us with a method of quantifying diffusion taking place in a brain due to the water molecules. Generally, two types of diffusion taking place in the brain are isotropic diffusion and anisotropic diffusion. In isotropic diffusion, there is equal diffusion of water molecules in all directions whereas, in anisotropic diffusion, the diffusion of water molecules varies in a different direction. In places near white matter fibers, most of the water molecules undergo anisotropic diffusion. Most of the water molecules diffuse in the direction parallel to the white matter fibers. The DTI method with the help of DWI generates three eigenvectors in each voxel.
The diffusion tensor models the diffusion signal as:

$$\frac{S\left(g,b\right)}{S_0} = e^{-bg^T D_g} . \qquad (1)$$

where $g$ .is a unit vector in 3 spaces indicating the direction of measurement and $b$ .denotes the magnetic field. $S\left(g,b\right)$ .represents the diffusion-weighted signal measured and $S_0$ .is the signal measured when there is no diffusion encoding due to magnetic field. $D$ .denotes a $3 * 3$ .covariance matrix of diffusivity along the three spatial dimensions.

$$D = \begin{bmatrix} D_{xx} & D_{xy} & D_{xz} \\ D_{yx} & D_{yy} & D_{zz} \\ D_{zx} & D_{zy} & D_{zz} \end{bmatrix}. \quad (2)$$

Metrics such as fractional anisotropy (FA) indicates the density of packing of fibers in a voxel and the amount of myelin wrapping the axons of the neurons. FA for a voxel depends on the three eigenvectors reconstructed from the voxel. Let the three eigenvectors be represented by $\lambda_1, \lambda_2, \lambda_3$. FA for each voxel is calculated as follows:

$$FA = \sqrt{\frac{1}{2} \frac{\left(\lambda_1 - \lambda_2\right)^2 + \left(\lambda_1 - \lambda_3\right)^2 + \left(\lambda_2 - \lambda_3\right)^2}{\lambda_1^2 + \lambda_2^2 + \lambda_3^2}}. \quad (3)$$

The range of fractional anisotropy is between zero and one that describes the degree of anisotropy of a diffusion process. FA of values zero means that diffusion is isotropic, which means that the diffusion of water molecules is the same in all directions as it is unrestricted. A value of one means that majority of the diffusion is taking place dominantly in one direction. FA can help us understand the fiber density, axonal diameter, and myelination in white matter.

The eigenvectors can help us to describe the shape of the diffusion profile in a voxel. If the three eigenvectors are equal in magnitude then it is spherical in shape or if one of the eigenvector is significantly large than the other two then the shape is ellipsoid. The ellipsoid shape represents that the diffusion is restricted in two directions and predominantly the water molecules flow in one direction. If the shape of the diffusion profile is spherical then the type of diffusion is isotropic and if the shape is ellipsoid then diffusion is anisotropic. The dominant vector in anisotropic diffusion is referred to as the principal eigenvector. All the principal eigenvectors when connected forms the white matter fibers.

*Figure 5. Coronal view of the brain with principal eigen vectors in each voxel*

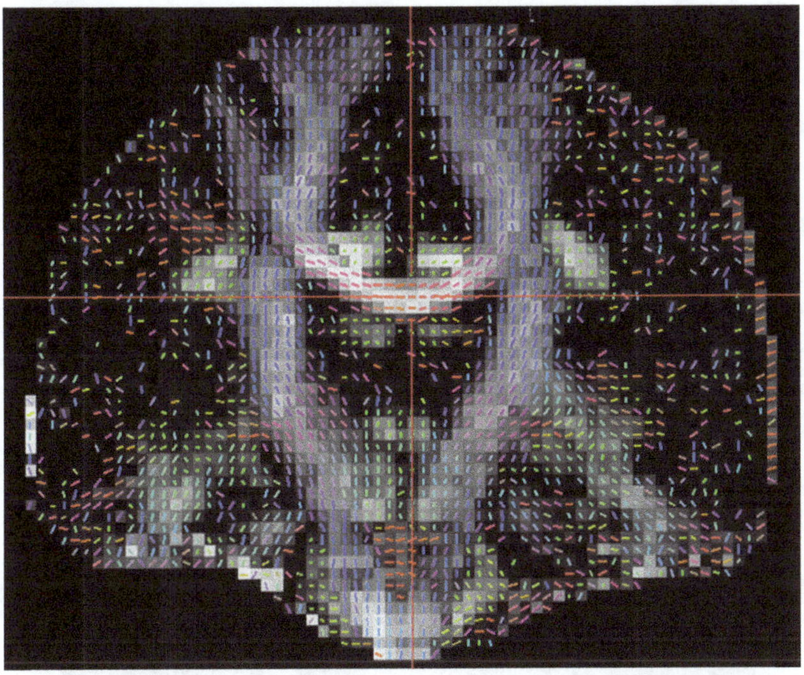

*Figure 6. Sagittal view of the brain*

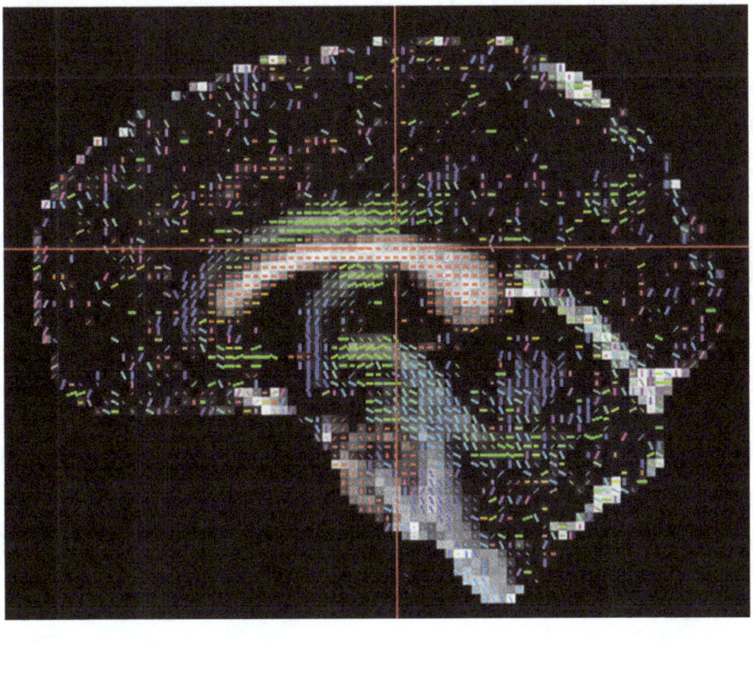

*Figure 7. Axial view of the brain*

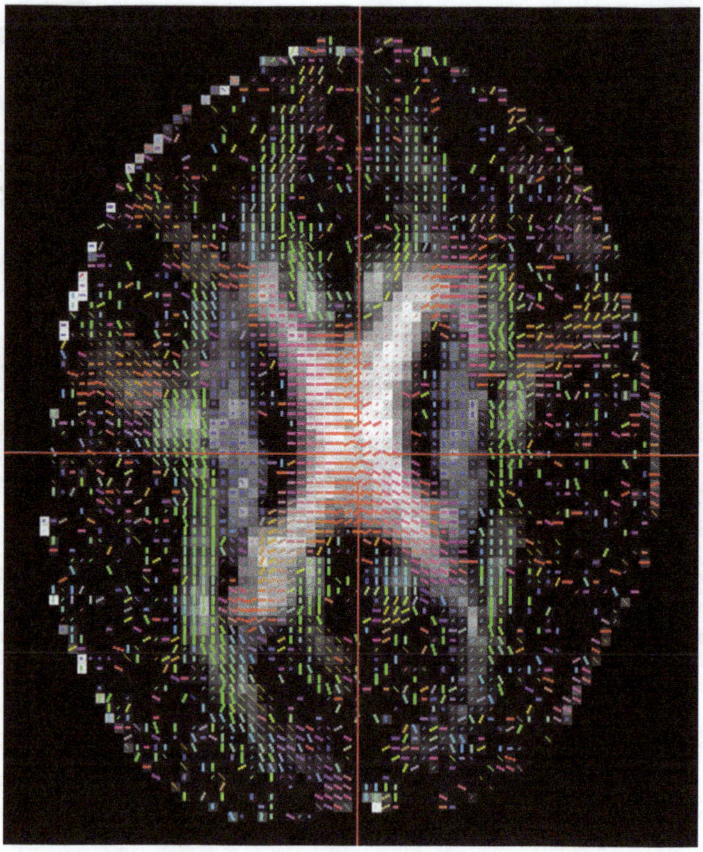

According to (Lauren & Carl, 2011) one of the most common method to reconstruct neurons is streamline tractography. This method is based on the approach of successively stepping in the direction of the principal eigenvectors which represents the direction of the neurons. Computational methods such as Euler's method and Runge-Kutta can be used to perform tractography of white matter fibers. However, there are certain limitations associated with the DTI model as the tracking is done only in one direction of the principal eigenvector. DTI model does not perform well when there is more than one fiber present in a voxel. The model can't track well when there are kissing or crossing fibers. As a result, fewer fibers are reconstructed. In the above images we can see some areas, where there are no principal eigenvectors visible especially in the top right corner of the sagittal view of the brain. Probabilistic-based tractography algorithms can address the limitation of the deterministic algorithm. It can help to reconstruct fibers in voxel where more than one fiber population is present.

## PROBABILISTIC TRACTOGRAPHY ALGORITHM

Similar to deterministic fiber tracking, the probabilistic approach follows the white matter fibers in a step-wise fashion and propagates streamlines based on the local orientations reconstructed at each voxel. The distribution of tracking directions at each point can be represented with the help of a probability mass function, where the tracking directions are restricted to the discrete number of points on a sphere. According to (Berman JI et.al, 2008) Q-ball fiber tracking algorithm is capable of tracking multiple intravoxel fiber populations. For instance, the algorithm has demonstrated that it is capable of following the corticospinal tract and corpus callosum through regions of crossing fibers in the centrum semiovale. Probabilistic tractography helps to address some of the limitations faced by DTI model such as kissing or fanning fibers. This model helps to predict more than one fiber in the same voxel. But the issue here it that this type of algorithm has certain criteria which need to be satisfied and is not data-driven.

## DEEP LEARNING ALGORITHM

(Itay and Tammy, 2019) developed a sequence-based model which is a type of RNN based on a gated recurrent unit (Recurrent Neural Networks) to predict the fiber orientation distribution function in each voxel. They used a discrete representation of the fiber orientation distribution function (fODF) by sampling a unit sphere. Models of fODF using unit sphere are available in the python library DIPY developed exclusively for working on diffusion MRI images.

Deep tract utilized the sequential nature of the diffusion MRI values to yield a conditional estimation of the fODF at a particular voxel. During the training of the model, the predicted conditional fiber orientation distribution function at a voxel is compared to the true orientation. When the model is given an initial seed point, the corresponding vector to the point is fed into the network. The network predicts the conditional fiber orientation distribution function based on the given vector. The proposed network consists of five stacked GRU layers, with each layer containing 1000 units. The loss function used was mean cross-entropy between the predicted conditional fiber orientation distribution function and the true labels along the fibers.

After the deep tract model has been trained tracking of the white matter fibers was done. The advantage of the deep tract is that both deterministic and probabilistic tracking can be done at the same time.

## FUTURE RESEARCH DIRECTIONS – REINFORCEMENT ALGORITHM

Currently, in clinical applications, DTI (Diffusion Tensor Imaging) is used which is a type of deterministic algorithm. As we have seen above deterministic algorithm does not perform better when there is more than one fiber is present in a voxel. Moreover, deterministic and tractography algorithm is model-based and lays down specific rules which need to be followed thus having restrictions on the data acquisition. Algorithms based on deep learning and machine learning require more data to be acquired for the tracking of fibers to be accurate. As a result, more research needs to be done on the data acquisition of the diffusion MRI to improve the performance of the algorithm.

Recently, Deep Mind developed a deep learning-based algorithm called reinforcement algorithm which defeated the game AlphaGo. (Antonie, Christian, Maxime & Pierre, 2020) used reinforcement algorithm to do tractography. This algorithm does not depend on ground truth and uses the concept of reward to track fibers. Reinforcement algorithm can reduce the requirement of more data and can work seamlessly in tracking fiber bundles. Research can be done on how to bring reinforcement-based algorithms into clinical applications.

## CONCLUSION

To aid neurosurgeons while operating on patients knowing the neural connectivity will be helpful for them. Our brain is filled with many fluids predominantly water which undergoes the phenomena of diffusion. The diffusion of water molecules can be captured using diffusion MRI which can be used to reconstruct the white matter fiber bundles such as corpus callosum and so on. Generally, the majority of the tractography algorithms can be divided as deterministic and probabilistic. Some of the limitations of the deterministic algorithm such as kissing or crossing are addressed with the help of probabilistic algorithms. Both deterministic and probabilistic algorithms are based on assumptions, hence impose specific criteria which need to be followed. To make tractography more data-driven researchers started using machine learning and deep learning algorithms. Traditional deep learning models are data-driven whereas currently developed algorithms such as reinforcement learning are based on reward rather than the ground truth data.

## REFERENCES

Benou, I., & Raviv, T. R. (2019). DeepTract: A probabilistic Deep Learning Framework for White Matter FIber Tractography. *International Conference on medical image computing and computer-based intervention.* 10.1007/978-3-030-32248-9_70

Berman, J. I., Chung, S., Mukherjee, P., Hess, C. P., Han, E. T., & Henry, R. G. (2008). Probabilistic streamline q-ball tractography using the residual bootstrap. *NeuroImage, 39*(1), 215–222. Advance online publication. doi:10.1016/j.neuroimage.2007.08.021 PMID:17911030

Lauren, J. (2011). An introduction to diffusion tensor image analysis. *Neurosurgery Clinics of North America, 22*(2), 185–196. doi:10.1016/j.nec.2010.12.004 PMID:21435570

Neher, Gotz, Norajitra, Weber, & Maier-Hein. (2015). A machine learning based approach to Fiber tractography using Classifier Voting. Medical Image Computing and Computer-Assisted Intervention (MICCAI), 45–52.

Poulin, P., Cote, M.-A., Houde, J.-C., Petit, L., Neher, P. F., Maier-Hein, K. H., Larochelle, H., & Descoteaux, M. (2017). Learn to track: Deep learning for tractography. Medical Image Computing and Computer-Assisted Intervention (MICCAI), 540–547. doi:10.1101/146688

Theberge, A., Desrosiers, C., Descoteaux, M., & Jodoin, P.-M. (2020). Track-To-Learn: A general framework for tractography with deep reinforcement learning. doi:10.1101/2020.11.16.385229

## ADDITIONAL READING

AlphaGo. (n.d.). https://deepmind.com/research/case-studies/alphago-the-story-so-far

DIPY. (n.d.). https://dipy.org/documentation/1.4.0./examples_index/#diffusion-tensor-imaging

ISMRM 2015 Tractography challenge. (n.d.). TRACTOMETER BY SCIL. http://www.tractometer.org/ismrm_2015_challenge/data

## KEY TERMS AND DEFINITIONS

**Diffusion:** Movement of water molecules from an area of higher concentration to lower concentration.

**dMRI:** dMRI stands for diffusion Magnetic Resonance Imaging which is a popular method used by neuroscientists to uncover unique information about the neural connections within the brain. Tractography - A 3D modelling technique used to represent the white matter fibers inside the brain.

**DTI:** DTI refers to diffusion tensor imaging which is a type of deterministic method predominantly used in clinical applications.

**DWI:** DWI refers to diffusion weighted image which is captured with the help of dMRI. Captures the diffusion information of the water molecules inside the brain.

**fODF:** Defined for all fibers in the unit. Indicates the probability of a fiber in a given voxel.

**Voxel:** A 3D unit that embeds the signal in brain scans.

**White Matter Fibers:** Areas of the central nervous system that are made up of myelinated axons. Also referred to as tracts.

# Chapter 7
# Cerebral Venous Thrombosis

**Ozcan Demetgul**
*Hatay Goverment Hospital, Turkey*

## ABSTRACT

*Cerebral venous thrombosis (CVT) is a rare condition involving various symptoms that is mainly seen in younger adults. The most commonly involved are the superior sagittal sinus, lateral sinus, and simoid sinus. About 1% of all ischemic strokes are considered CVT. It is seen 3-fold more in young women. The incidence was estimated to be approx. 1/1000000. Cerebral venous sinuses are superficial and deep spaces, and they have vital functions. There are many symptoms. The most common complaint is headache (89%). Studies reported many different clinical symptoms. Neurological signs including motor and sensorial losses, impaired consciousness, speech disorder, epileptic seizures, visual problems (hemianopia, nystagmus, diplopia, and papilledema), and cranial nerve signs may be seen. The diagnosis is made primarily by suspecting the clinical condition and radiological presentation of thrombosis. The most basic diagnostic method is cranial imaging. Anticoagulants are the main method of treatment. The prognosis has improved over the last years thanks to early diagnosis.*

## INTRODUCTION

Cerebral venous thrombosis (CVT) is a rare condition involving various symptoms which is mainly seen in younger adults (Masuhr et al., 2004). About 1% of all ischemic strokes are considered as CVT. It is seen 3-fold more in young women. The incidence was estimated to be approx. 1/1000000 (Bousser & Ferro, 2007).

DOI: 10.4018/978-1-7998-7433-1.ch007

## ANATOMY

Cerebral venous sinuses are superficial and deep spaces formed by dura mater. They have vital functions. Cerebral veins and sinuses do not involve a valve system. These two groups of venous system are interconnected via multiple intracerebral and extracerebral anastomotic structures. Eventually, they merge to form internal jugular vein. While superficial system basically consists of superior sagittal sinus and cortical veins; deep system includes transverse sinus, vein of Galen, straight sinus and sigmoid sinus (Dere, 2000).

## CLINICAL PRESENTATION

Although there are many symptoms, the most common complaint is headache (89%). Studies reported many different clinical symptoms. Neurological signs including motor and sensorial losses, impaired consciousness, speech disorder, epileptic seizures, visual problems (hemianopia, nystagmus, diplopia, and papilledema), and cranial nerve signs may be seen (Duman et al., 2017; Nagaraja & Sarma, 2002).

## ETIOLOGY

*Table 1. The most common risk factors for CVT*

| Pregnancy and puerperium |
|---|
| Oral contraception and hormone drugs |
| Antiphospholipid syndrome |
| Hereditary thrombophilia |
| Anemia |
| Female obesity |
| Cancer |
| Infections |
| Trauma |
| Iatrogenic |
| Idiopathic |

Contrary to ischemic strokes, there is an underlying cause in most of the cases (Table 1). Common etiologies include pregnancy, puerperium, oral contraceptive use (OCPs), coagulation disorders, infection, brain tumors and connective tissue disorders (Duman et al., 2017). It is common in women of childbearing age with almost half of the ischemic strokes during pregnancy and lactation are cerebral venous strokes (Cantu-Brito et al., 2011).

## PATHOPHYSIOLOGY

While the pathophysiology is complex, there are usually 2 important mechanisms. First, there is increased intracranial pressure and parenchymal ischemia. Venous thrombosis reduces perfusion pressure, blood-brain-barrier gets impaired and vasogenic edema occurs leading to hemorrhages, brain perfusion gets gradually impaired resulting in brain damage with cytotoxic edema. Moreover, thrombosis in the sinuses decreases the passage of cerebrospinal fluid and increases the intracranial pressure(Schaller & Graf, 2004).

## THE DIAGNOSIS OF CVT

The diagnosis is made primarily by suspecting the clinical condition and radiological presentation of thrombosis. The most basic diagnostic method is cranial imaging. Superior sagittal and transverse sinus thrombosis are seen most frequently in examinations. There is no special serological marker to confirm the diagnosis, but it is recommended to look at the time of hemogram, biochemistry panel and prothrombin time. Although high D-dimer serum level supports the diagnosis, its normal occurrence does not rule out the diagnosis.

Computed tomography (CT) is a main widely used imaging method for ease of transportation. Non-contrast CT does not have much place in diagnosis. It is not sufficient to demonstrate sinus variations and ischemic changes. The most important finding is the hyperdense appearance of the thrombosed cortical vein or dural sinus, which can be seen in 1/3 of the cases. Thrombosis of the posterior part of the superior sagittal sinus leads to hyperdense appearance of the sinus in non-contrasted CT, and this finding is called the “delta sign”. In contrast-enhanced CT, there is no contrast in the middle of the sinus due to thrombus, but an all-round contrast rim can be seen. This finding is called “empty delta sign” due to contrast filling defect. (Boukobza et al., 2007).

Magnetic resonance venography (MR-V) and magnetic resonance imaging (MRI) can also be classified as sensitive instruments for diagnosing CVT . These tools have the advantage of being more susceptible to motion and low-flow image artefacts, more sensitive for identifying alternative pathologies, as well as decreasing exposure to ionising radiation. Appearance may change depending on thrombus time(Saposnik et al., 2011; Tentschert et al., 2005).

## LOCATION

The most commonly involved are the superior sagittal sinus, lateral sinus and simoid sinus. The most common thrombosis sites can be classified as the transverse sinus (60.5%) and the superior sagittal sinus (65%) (Sassi et al., 2017). A multicenter study reported that deep venous system involvement was observed in 10.9% of CVT patients, cortical vein involvement was found in 17.1% of CVT patients and jugular vein involvement was detected in 11.9% of the patients. Only 0.3 and 1.3% of the patients exhibited the cerebellum and cavernous sinus involvement, respectively (Ferro et al., 2004).

## TREATMENT

Anticoagulants are the main method of treatment. Unlike arterial strokes, antiaggregants are not effective. Furthermore, as a general rule for treatment in medicine, treatment for the underlying cause, if detected, is absolutely necessary.

The general management approach consists of: 1) Anticoagulant therapy, 2) Decompressive surgery and 3) Endovascular treatment

### 1. Anticoagulant Therapy

Heparin is the touchstone of the treatment with low-molecular-weight heparin being regarded as superior than unfractionated heparin(Coutinho et al., 2010).

Today, the immediate initiation of anticoagulant treatment is recommended, and in addition to anticoagulation therapy, underlying etiological variables should analyzed and managed appropriately. Measures for reversible risk factors such as discontinuing prothrombic drugs, rehydration and antibiotics should be considered immediately(Saposnik et al., 2011).

Recanalization takes 3 to 12 months. After the acute period, it is recommended to continue oral anticoagulant treatment for 6 more months on average. Recanalization is also influenced by the affected anatomic region. Recanalization rate in transverse sinus is lower than the other sinuses (Baumgartner et al., 2003).

## 2. Decompressive Surgery

It is the treatment method used in case of increased intracranial pressure. Large hemorrhages seen in rare patient groups cause increased intracranial pressure leading to herniation. Surgical intervention is performed in these patients to reduce mortality(Théaudin et al., 2010).

## 3. Endovascular Treatment (ET)

Interventional procedures may be applied if medical treatment fails or is contraindicated. There is no clear indication or timing information. Endovascular treatment methods mainly include pharmacological and mechanical thrombectomy, balloon angioplasty, and Penumbra aspiration(Gulati et al., 2014).

# PROGNOSIS

The prognosis has improved over the last years thanks to early diagnosis. In a larger study, mortality rate was observed to be 4.3%. The most important factors for mortality include hemorrhages, edema and herniation(Ferro et al., 2004).

# REFERENCES

Baumgartner, R. W., Studer, A., Arnold, M., & Georgiadis, D. (2003). Recanalisation of cerebral venous thrombosis. *Journal of Neurology, Neurosurgery, and Psychiatry*, *74*(4), 459–461. doi:10.1136/jnnp.74.4.459 PMID:12640063

Boukobza, M., Crassard, I., & Bousser, M. G. (2007). When the "dense triangle" in dural sinus thrombosis is round. *Neurology*, *69*(8), 808. doi:10.1212/01.wnl.0000275548.60964.bf PMID:17709717

Bousser, M. G., & Ferro, J. M. (2007). Cerebral venous thrombosis: An update. *Lancet Neurology*, *6*(2), 162–170. doi:10.1016/S1474-4422(07)70029-7 PMID:17239803

Cantu-Brito, C., Arauz, A., Aburto, Y., Barinagarrementeria, F., Ruiz-Sandoval, J. L., & Baizabal-Carvallo, J. F. (2011). Cerebrovascular complications during pregnancy and postpartum: Clinical and prognosis observations in 240 Hispanic women. *European Journal of Neurology*, *18*(6), 819–825. doi:10.1111/j.1468-1331.2010.03259.x PMID:21143339

Coutinho, J. M., Ferro, J. M., Canhão, P., Barinagarrementeria, F., Bousser, M. G., & Stam, J. (2010). Unfractionated or low-molecular weight heparin for the treatment of cerebral venous thrombosis. *Stroke*, *41*(11), 2575–2580. doi:10.1161/STROKEAHA.110.588822 PMID:20930161

Dere, F. (2000). *Nöroanatomi Atlası ve Ders Kitabı, üçüncü baskı*. Nobel Yayınları.

Duman, T., Uluduz, D., Midi, I., Bektas, H., Kablan, Y., Goksel, B. K., Milanlioglu, A., Necioglu Orken, D., Aluclu, U., Colakoglu, S., Tüfekci, A., Bakar, M., Nazliel, B., Tascilar, N., Goksan, B., Kozak, H. H., Demir, S., Misirli, C. H., Kucukoglu, H., ... Uzuner, N. (2017). A Multicenter Study of 1144 Patients with Cerebral Venous Thrombosis: The VENOST Study. *Journal of Stroke and Cerebrovascular Diseases*, *26*(8), 1848–1857. doi:10.1016/j.jstrokecerebrovasdis.2017.04.020 PMID:28583818

Ferro, J. M., Canhão, P., Stam, J., Bousser, M. G., & Barinagarrementeria, F. (2004). Prognosis of cerebral vein and dural sinus thrombosis: Results of the International Study on Cerebral Vein and Dural Sinus Thrombosis (ISCVT). *Stroke*, *35*(3), 664–670. doi:10.1161/01.STR.0000117571.76197.26 PMID:14976332

Gulati, D., Strbian, D., & Sundararajan, S. (2014). Cerebral venous thrombosis: Diagnosis and management. *Stroke*, *45*(2), e16–e18. doi:10.1161/STROKEAHA.113.003964 PMID:24357654

Masuhr, F., Mehraein, S., & Einhäupl, K. (2004). Cerebral venous and sinus thrombosis. *Journal of Neurology*, *251*(1), 11–23. doi:10.100700415-004-0321-7 PMID:14999484

Nagaraja, D., & Sarma, G. R. (2002). Treatment of cerebral sinus/venous thrombosis. *Neurology India*, *50*(2), 114–116. PMID:12134170

Saposnik, G., Barinagarrementeria, F., Brown, R. D. Jr, Bushnell, C. D., Cucchiara, B., Cushman, M., deVeber, G., Ferro, J. M., & Tsai, F. Y. (2011). Diagnosis and management of cerebral venous thrombosis: A statement for healthcare professionals from the American Heart Association/American Stroke Association. *Stroke*, *42*(4), 1158–1192. doi:10.1161/STR.0b013e31820a8364 PMID:21293023

Sassi, Touati, Baccouche, Drissi, Romdhane, & Hentati. (2017). Cerebral Venous Thrombosis: A Tunisian Monocenter Study on 160 Patients. *Clinical and Applied Thrombosis/Hemostasis, 23*(8), 1005-9.

Schaller, B., & Graf, R. (2004). Cerebral venous infarction: The pathophysiological concept. *Cerebrovascular Diseases (Basel, Switzerland)*, *18*(3), 179–188. doi:10.1159/000079939 PMID:15273432

Tentschert, S., Wimmer, R., Greisenegger, S., Lang, W., & Lalouschek, W. (2005). Headache at stroke onset in 2196 patients with ischemic stroke or transient ischemic attack. *Stroke*, *36*(2), e1–e3. doi:10.1161/01.STR.0000151360.03567.2b PMID:15604424

Théaudin, M., Crassard, I., Bresson, D., Saliou, G., Favrole, P., Vahedi, K., Denier, C., & Bousser, M.-G. (2010). Should decompressive surgery be performed in malignant cerebral venous thrombosis? A series of 12 patients. *Stroke*, *41*(4), 727–731. doi:10.1161/STROKEAHA.109.572909 PMID:20185780

# Chapter 8
# Clinical Features of Stroke

**Alper Uysal**
*Hatay Training and Research Hospital, Turkey*

## ABSTRACT

*Stroke is a clinical condition that causes neurological dysfunction due to focal infarction or haemorrhage in the brain, spinal cord, or retina. These clinical features may take 24 hours or more and result in death. Stroke is one of the leading causes of disability and death. With the prolongation of life in societies, stroke and stroke-related risk factors become more and more important. Age, gender, race, heredity, ethnicity, hypertension, atrial fibrillation, diabetes, hyperlipidemia, smoking, transient ischemic attack, and physical inactivity are risk factors of stroke. Signs and symptoms of stroke vary according to occluded vessel. Mental dysfunction, speech and language disorders, motor and sensory impairment may occur as a result of stroke.*

## INTRODUCTION

A stroke, which is the second leading cause of death and disability worldwide, is an acute irreversible focal neurological deficit syndrome that occurs as a result of vascular damage to the central nervous system and approximately 16.9 million people worldwide suffer a stroke each year (Mehndiratta, Smith, & Worrall, 2015; Murphy & Werring, 2020). This article sheds light on the pathophysiology of the stroke, its clinical manifestations associated with vascular lesions and post-stroke neurologic deficits.

Approximately 85% of strokes are of ischemic origin (associated with thrombus or embolism), while 15% are of intracerebral hemorrhagic origin. Intracerebral hemorrhage may be deep-seated or lobar. About 80% of hemorrhages are associated

DOI: 10.4018/978-1-7998-7433-1.ch008

with cerebral microvascular diseases (deep perforator arteriopathy, cerebral amyloid angiopathy), while 20% are associated with macrovascular lesions (arteriovenous malformations, aneurysms) or cerebral venous thrombosis (Murphy & Werring, 2020). Cerebral ischemia and hemorrhage can be differentiated by brain imaging methods. Several neurological deficits develop depending on the cerebral area perfused by the affected vascular structure. To determine the relationship between the vascular lesion and clinical picture, it is necessary to know the anatomy of the cerebral artery and the representative localization of different functions in the brain (Markus, 2008).

The cerebral arterial circulation is divided into two parts; the anterior and posterior circulation. There are 2 main arteries responsible for the blood supply to the brain; the internal carotid arteries and the vertebral arteries. While 80% of the cerebral blood flow is obtained from the carotid system, 20% of it is obtained from the vertebrobasilar system. The connection between these two systems is provided by the Willis polygon. The major arteries of the carotid system that supply the anterior parts of the brain are the internal carotid artery, middle cerebral artery and anterior cerebral artery. The main arteries of the vertebrobasilar system that supply the posterior brain regions are the vertebral arteries, basilar artery, cerebellar artery and posterior cerebral artery (Durlanık, 2016). As a result of cerebral vascular damage, blood circulation in the brain decreases and then neuronal cellular damage and neuronal loss develop (Frizzell, 2005).

## PATHOPHYSIOLOGY AND CLINICAL VIEWS

### Transient Ischemic Attack

Transient ischemic attack (TIA), as a classic definition, is a neurological deficit that causes neurological symptoms and signs as a result of focal ischemia in the brain and retina but does not cause infarction and lasts less than 24 hours. However, recent studies evaluating cognition in patients with TIA have shown that impairments in executive function did not improve on day 7 after TIA, but there are limited studies examining this condition. Thanks to current developments in neuroimaging methods, persistent microvascular tissue injury has been demonstrated in the brain after TIA. Small deficits in memory, attention and problem-solving skills can negatively affect a person in everyday life (Ganzer, Barnes, Uphold, & Jacobs, 2016). Recent studies have shown that 30–50% of patients have neuronal death in their cerebral tissue in diffusion MR, even if symptoms improve after TIA (Easton et al., 2009). Temporary visual loss may develop as a result of obstruction of the retinal artery with microembolism from the atherosclerotic plaques in the internal carotid aorta (Harvey,

Roth, & Yu, 2007). TIA is an important risk predictor for ischemic stroke. The risk of developing stroke increases significantly after TIA. This risk is highest in the first 48 hours and it is important for more than 3 months. Nearly 11% of people suffer a stroke in the first 90 days after TIA (Crowfoot, van der Riet, & Maguire, 2016).

## Cerebral Thrombosis

Cerebral thrombosis accounts for about 60% of all stroke cases and develops due to atherosclerosis and collateral circulatory failure. With regard to cerebrovascular thrombosis and associated with the severity and duration of ischemia, a TIA or minor stroke that does not cause disability or a major stroke that results in severe deficits and functional disability may occur. Inactivity or sleep creates susceptibility to thrombosis. As a result of cerebral oedema and metabolic changes in brain tissue, the patient's clinical symptoms often worsen in the first hours and days after stroke. Recovery usually begins at the end of the first week.

## Cerebral Embolism

Cerebral embolism accounts for about 20% of all stroke cases (Çevikol & Çakcı, 2015). The most common cause of cerebral embolism is a cardiac mural thrombus, but an embolus can originate from the internal carotid aorta or ascending aorta. The main heart diseases leading to cerebral embolism are cardiac arrhythmia, ischemic heart disease, valve disease and atrial septal anomalies (Pare & Kahn, 2012). Due to their large size, emboli are often more common in the irrigation area of the middle cerebral artery (MCA). It can also make superficial, massive, single large striatocapsular or multiple lesions. Global aphasia or Wernicke's aphasia may occur without hemiparesis. It has a sudden start and the ischemic infarction can transform into a haemorrhagic infarction or early recanalization may develop in the clogged vessel. As a result, the clinical picture may be variable. There is a formal indication for long-term anticoagulant therapy in patients with a history of TIA or stroke and atrial fibrillation (Ferro, 2003).

## Lacunar Stroke

Lacunar stroke occurs as a result of occlusion of the deep penetrating branches of the cerebral artery. Clinical findings vary according to the area irrigated by the clogged penetrating artery. Lacunar infarcts may present with pure motor hemiparesis, ataxic hemiparesis, dysarthria-clumsy hand syndrome, pure sensory stroke or sensorimotor stroke (Arboix & Martí-Vilalta, 2009). Hypertension, smoking, atherosclerosis and diabetes mellitus are included in the aetiologies of lacunar infarction. There is usually a history of TIA and gradual onset before a lacunar stroke. When lacunar infarcts are evaluated in terms of prognosis, recovery rates are high after lacunar infarcts and the risk of stroke recurrence and death is low (Donnan & Norrving, 2008).

## Intracerebral Haemorrhage

Intracerebral bleeding develops in hypertensive patients due to rupture of microaneurysms in the deep penetrating arteries. Microvascular changes such as lipohyalinosis and Charcot-Bouchard aneurysms can cause hypertensive bleeding. As a result of bleeding, increased intracranial pressure, ventricular compression and cerebral herniation may develop. Acute mortality is high, but in survivors, neurological deficits resolve rapidly within the first 2 or 3 months after bleeding. About half of intracranial haemorrhages occur in the putamen and cerebral white matter. Bleeding lasts less than 1–2 hours and acute symptoms appear during this time period. Neurological worsening in the late period is associated with post-haemorrhagic oedema or re-bleeding (Harvey et al., 2007). Haemorrhages taking place in the basal ganglia and thalamus are called deep bleeding and bleeding in the cerebral hemispheres is called lobar bleeding. Haemorrhages in the putamen constitute 50–60% of intraparenchymal haemorrhages (Pare & Kahn, 2012).

## Subarachnoid Haemorrhage

Subarachnoid haemorrhage (SAH) is bleeding that takes places in the dural space around the brain and the most widespread reason is rupture of an aneurysm or arteriovenous malformation (AVM). If a tear occurs in the vessel wall, blood flowing into the subarachnoid space irritates the dura and causes a terrible headache. Acute loss of consciousness is common. Focal neurological deficits can occur. Coma is common and about 1 in 3 of patients die instantly. Patients with aneurysms that have not received surgery have a 30% risk of bleeding again within the first month after SAH. The risk of bleeding gradually decreases and the long-term risk of bleeding is up to 3% per year. SAH may also develop as a result of AVM bleeding. Generally, congenital AVMs tend to bleed in the 2nd and 3rd decades of life. The risk of bleeding is 40–50% throughout life (Harvey et al., 2007).

# STROKE CLINICAL VIEWS RELATED TO VASCULAR LESIONS

## Internal Carotid Artery Syndrome

The carotid system and the vertebrobasilar system are responsible for the blood supply of the brain. These two systems are connected by the polygon of Willis. Internal carotid artery (ICA) syndrome occurs due to embolism or decreased blood flow. Both cerebral and retinal ischemia are observed in ICA lesions. If the polygon of Willis is intact, if the collaterals that provide the connection between the external carotid artery and ICA are not obstructed, neurological findings will not develop. ICA syndrome presents as a TIA or stroke and occurs in a small number of patients. TIA appears as transient monocular blindness (amaurosis fugax) and hemispheric TIA. It can occur in situations that tend toward hypotension, such as rapid standing. ICA syndrome is often associated with a lesion of the MCA or its branches. There are clinical signs and symptoms of an ICA lesion, such as contralateral motor deficit, contralateral sensory deficit, contralateral hemianopsia, conjugate eye deviation to the lesion side and higher cortical dysfunction (aphasia in the dominant hemisphere and neglect in the non-dominant hemisphere) (Durlanık, 2016).

## Middle Cerebral Artery Syndrome

While a MCA lesion in the dominant hemisphere may cause global aphasia, a MCA lesion in the non-dominant hemisphere may cause hemispatial neglect. The speech centre is in the left hemisphere in 70–80% of people with left-hand dominant use and in most of the right-hand dominant users. Motor and sensory losses that affect the face, upper limb and, to a lesser extent, the lower limb may develop on the contralateral side of the infarct. There may be loss of vision in the form of homogeneous hemianopia and conjugated eye aberration, which indicates the side of the infarct. The MCA has superior and inferior branches.

The superior branch irrigates Broca's area of the frontal lobe in the dominant hemisphere and motor aphasia is seen in lesions of Broca's area. The inferior branch irrigates Wernicke's area in the temporal lobe of the dominant hemisphere and sensory aphasia is observed with its ischemia. In Broca's aphasia, speech stops or is poorly expressed, whereas in Wernicke's aphasia, speech is fluent. But since the patient uses false and nonsense words, it is difficult to understand what he/she wants to say (Pare & Kahn, 2012).

## Anterior Cerebral Artery Syndrome

A lesion of the anterior cerebral artery can cause contralateral motor and sensory losses that affect the lower limb more than the upper limb, because ischemia develops in the paracentral lobule of the interhemispheric cortex. One-sided low foot, which requires long-term orthotic use, may develop. The face and tongue are often not affected. Incontinence and primitive frontal lobe reflexes, such as sucking and gripping may occur (Harvey et al., 2007; Pare & Kahn, 2012).

## Posterior Cerebral Artery Syndrome

Colour agnosia and associated visual agnosia are typical deficits in left posterior cerebral artery syndrome (PCA) lesions. Hemianopic alexia and alexia without agraphia can occur. Even though facial agnosia is more widespread in bilateral lesions, it may also develop in unilateral lesions. Bilateral lesions may produce pronounced visual agnosia or cortical blindness. Denial and confabulation usually occur. Memory defects are observed in 25% of patients, especially in lesions of the mesial part of the temporal lobe. Memory loss is more common in bilateral or left sided temporal lobe lesions (Ferro, 2001).

## Superior Cerebellar Artery Syndrome

The main symptoms of superior cerebellar artery occlusion are neurological deficits such as ipsilateral cerebellar ataxia, ipsilateral Horner's syndrome, loss of contralateral pain and temperature sensation and contralateral involvement of the 4th cranial nerve. Dysarthria, tremor, diplopia, vomiting, tetraplegia and coma can also occur.

## Anterior Inferior Cerebellar Artery Syndrome

This artery irrigates the lateral area of the lower sections of the pons and the flocculus and the middle cerebellar peduncle of the cerebellum and when obstructed, coma; 5th, 7th and 8th cranial nerve lesions; vertigo; tinnitus; dysarthria; Horner's syndrome; contralateral pain and temperature sensory losses; and pure vestibular syndrome may occur.

## Posterior Inferior Cerebellar Artery Syndrome

In the cerebellum, the infarct takes place most commonly in the area that is irrigated by the posterior inferior cerebellar artery (PICA) (Pare & Kahn, 2012). Lesions of the PICA present with vertigo, vomiting, ataxia, headache, delayed coma (pseudotumor form), Wallenberg syndrome, dysmetria, ipsilateral lateropulsion and ataxia (Amarenco, 2001).

## Vertebrobasilar Artery Syndrome

Emboli (30–40%) are the most common reasons for the aetiology of vertebrobasilar artery syndrome (VBAS). Thrombosis (15–35%) and vertebral artery dissection (about 5%) are other causes. However, since multiple vascular pathologies are more common in elderly patients, local thrombosis is ranked first in aetiology of VBAS in elderly patients. In young people, the most common cause is embolism and multiple vascular pathologies are rarely seen. While the clinical embolism develops suddenly, the clinical thrombosis develops slowly and progressively; TIA is frequently seen before the stroke. VBAS clinically varies according to the region being affected. The most common symptoms at first admission are nausea, vomiting, dizziness, vertigo, headache or neck pain. Loss of strength, hemiparesis, ataxia, diplopia, pupillary abnormalities and speech disorders may also develop in VBAS. Intracranial vertebral artery occlusion can also occur as Wallenberg syndrome, which is also known as lateral medullary syndrome. Clinical features of this syndrome may include loss of pain and heat sensation in the ipsilateral face and contralateral trunk and limbs, ataxia (extremity), Horner's syndrome, dysphagia and hoarse voice. Locked-in syndrome, which presents as quadriplegia, anarthria and intact cognition due to pons infarction, is observed in basilar artery involvement. Top-of-the-basilar syndrome (coma, hemiataxia, hemiparesis and visual symptoms) may develop in distal basilar artery lesion (Schoen, Boysen, Warren, Chakravarthy, & Lotfipour, 2011).

## Cerebral Venous Thrombosis

Cerebral venous thrombosis (CVT) is the development of thrombosis in any of the venous sinuses in the brain. CVT is rarely involved in the aetiology of cerebral infarction compared to arterial occlusive diseases of the brain. CVT often appears as a headache initially and is sometimes accompanied by cranial nerve palsy. Seizures and/or coma may occur in patients with more severe involvement. Papilloedema is a common finding in these patients (Pare & Kahn, 2012).

# POST-STROKE NEUROLOGICAL DISORDERS

## Mental Dysfunction

Formerly, post-stroke dementia was considered to be vascular dementia. However, after it was realized that all stroke patients with deficits in their cognitive functions do not meet the dementia criteria, the term vascular cognitive disorder has been used instead of the term vascular dementia. Thus, other cognitive disorders not accompanied by dementia were included in this definition (Sun, Tan, & Yu, 2014). The most common type of dementia is Alzheimer's disease (AD), with a rate of about 50–70%. The second most common cause of dementia is vascular cognitive impairment (VCI), with a rate of about 15–25%. The coexistence of AD and VCI is estimated to be around 50% and this coexistence is named mixed dementia (Sahathevan, Brodtmann, & Donnan, 2012). The incidence of dementia increases by 10 times during the 1-year period post-stroke as compared with stroke-free people (Ukraintseva, Sloan, Arbeev, & Yashin, 2006). Approximately one-third of the patients develop dementia within 1 year after stroke; however, about 60% of those who do not meet the criteria for dementia have cognitive deficits (Hurford, Charidimou, Fox, Cipolotti, & Werring, 2013).

Cognition is a whole of many processes, such as attention, executive function, visual spatial ability, memory and language. Neuropsychological tests involving various areas are used to detect cognitive deficits (Cumming, Marshall, & Lazar, 2013) and among these tests, the Mini-Mental State Examination (MMSE) score is the gold standard in diagnosis. The MMSE score is calculated by giving points to place-time orientation, repetition of three words, attention, account, remembering three words and language. A score ≤ 24 points (out of a maximum total of 30 points) suggests cognitive impairment. The MMSE is poor in evaluation of frontal and executive functions. In the watch drawing test, the patient is asked to draw a watch and place it in the figures. Afterwards, he/she is asked to draw the scorpion and the winding hive according to the time he/she was told. This test is a short, cheap and easy to apply diagnostic method (Palsetia, Rao, Tiwari, Lodha, & De Sousa, 2018). The MMSE can be used to evaluate praxis, memory, orientation, language and constructive skills and the watch drawing test can be used to evaluate high cortical functions. The prefrontal system of the brain, which represents executive functions, can be evaluated with the watch drawing test (Román, 2003). Cognitive deficits such as aphasia, mutism, agraphia, acalculia, confusion, intellectual and buccal-facial apraxia are seen in left hemisphere MCA occlusion and visual-spatial disorders, visual-constructive disorders, neglect, aprosodia, anosognosia and pragmatic language are observed in right hemisphere MCA occlusion.

Colour agnosia, associative visual agnosia, facial agnosia, amnesia, alexia (hemianopic and pure) and Balint syndrome are the main cognitive deficiencies in PCA occlusions. Concentration, planning, initiation, flexibility and monitoring may also be affected in bilateral anterior cerebral artery occlusion. In subcortical infarcts involving thalamic infarcts, deficits associated with arousal, memory, initiation, attention, motivation and executive function develop. Caudate infarcts can cause problems related to problem solving, memory and attention. Subcortical lesions of the inferior genus of the internal capsule can cause memory loss and confusion. SAH associated with anterior communicating artery aneurysm rupture causes damage to the mesial frontal cortex. Confabulation, amnesia, personality changes and abulia may develop (Donovan et al., 2008).

Memory deficits occur after stroke in 20–50% of patients (Al-Qazzaz, Ali, Ahmad, Islam, & Mohamad, 2014). There is an extreme slowing in the processing of information after a stroke and this seriously affects patients' daily life activities (DLA). There is a more pronounced deficit in attention and executive functions compared to memory and language processing after stroke. In the first year after stroke, inadequate improvement in attention deficit was observed, while problems with language and memory improve. It was found that both visual and verbal memories were in better condition, while processing speed and executive function were poor 5 years after stroke (Cumming et al., 2013). Compensatory strategies, anodal transcranial direct current stimulation, repetitive transcranial magnetic stimulation and music listening are the treatment methods that can be effective in healing memory problems (Buğdaycı & Caf, 2016).

Unilateral spatial neglect (USI) is a cognitive deficit that usually develops due to stroke. The response of the patient to the sensory stimuli from the contralateral lesion was impaired. There are two subtypes: allocentric neglect and egocentric neglect. In egocentric (observer-based) neglect, the person ignores the stimuli on the contralesional side of the midline of his/her body. In allocentric (object-based) neglect, the person ignores the stimuli on the contralesional side of an object he/she focuses on (Demeyere & Gillebert, 2019).

According to Chechlacz et al, left allocentric neglect is observed in posterior cortical region infarcts of the right hemisphere, such as the right posterior superior temporal sulcus, angular, middle temporal and middle occipital gyrus. Left egocentric neglect is observed in more anterior cortical infarcts, such as right middle frontal, postcentral, supramarginal and superior temporal gyrus and the insula. Intraparietal sulcus and temporoparietal junction infarcts can cause both types of neglect (Chechlacz et al., 2010). The star cancellation test, pointing to objects located about the ward, food on a plate, line cancellation, coin selection, figure copying from the left and line bisection tests are the main tests used in the diagnosis of neglect (Stone et al., 1991). In their 2001 study, Ferber and Karnath found cancellation tests more sensitive than bisection tests in detecting neglect. In the star cancellation test, the patient is asked to mark all the small stars on an A4 sheet with 52 big stars, 56 small stars, 13 letters and 10 short words. Since two small stars in the centre are not included in the scoring, the maximum score is 54. There should be 27 little stars in each half. The number of small stars skipped on the left, right and in total should be recorded (Ferber & Karnath, 2001).

For treatment, it is aimed to modulate the spatial reference system by stimulating the sensory afferents with optokinetic stimulation, vestibular and galvanic stimulation and electrical stimulation of the neck muscles. These options are very successful in the treatment of neglect. However, the healing effect is short-lived. Mirror therapy has been shown to be very successful in experimental studies. Prism therapy, stimulation of arousal and non-invasive brain stimulation techniques, such as transcranial magnetic stimulation (TMS) and transcranial direct current stimulation (tDCS), are methods that can be used in therapy. It is aimed to provide interhemispheric balance with non-invasive brain stimulation. Studies on healthy subjects have shown that neglect can be controlled by reducing the activity of the left hemisphere (repeated low frequency TMS (< 1 Hz) and cathode tDCS) or increasing the activity of the right hemisphere (repeated high frequency TMS (> 5 Hz) and anodal tDCS) (Azouvi, Jacquin-Courtois, & Luauté, 2017).

Apraxia is a movement disorder in which a person cannot perform a voluntary activity, even though there is no loss of strength, sensation or coordination (Park, 2017). Apraxia occurs in 9.1–30% of patients after stroke. Its incidence is higher in left hemisphere lesions. Even though apraxia is frequently seen in left parietal lobe lesions, right parietal, temporal and frontal lobe and subcortical lesions may be associated with apraxia (Salter, Teasell, Foley, & Allen, 2013). Ideomotor, ideational and limb-kinetic apraxia are the most common forms of apraxia. In ideomotor apraxia, the patient explains how to do an activity, but when he/she is asked to perform the activity, he/she cannot do it. If a patient with ideational apraxia is given a glass and jug, the patient will name the glass and jug but cannot show how to put water in the glass. The patient cannot perform a series of sequential actions in the correct order to perform an activity. In limb-kinetic apraxia, coordinated hand and finger movements are done incorrectly or incompetently. Dressing apraxia, eyelid opening apraxia and walking apraxia are examples of task-specific apraxia (Park, 2017).

There is strong (Level 1a) evidence to support that strategy (compensatory) training is effective in curing stroke-related apraxia. Improvement in DLA is observed. There is strong (Level 1a) evidence that gesture training in ideomotor apraxia provides healing. In DLA, improvement can be observed for at least 2 months after treatment (Salter et al., 2013). Apraxia treatment is more effective if non-invasive brain stimulation techniques are combined with rehabilitation training. tDCS, single-pulse or rTMS, theta-burst stimulation (TBS) and paired associative stimulation

(PAS) are examples of non-invasive brain stimulation techniques. TBS, which is an example of magnetic stimulation, has an equal effect with shorter stimulation period compared to other stimulation techniques (Park, 2017). Donepezil and galantamine are cholinesterase inhibitors that may be used as first-line treatments for vascular dementia. Memantine (NMDA receptor antagonist), pentoxifylline, calcium channel blockers, nootropics and antiplatelet agents have a positive effect on cognitive function. The primary approach in treatment should be to control the risk factors of cerebrovascular disease, such as hypertension, diabetes mellitus, atherosclerosis, hyperlipidaemia and cardiogenic embolism (Román, 2003).

## Speech and Language Disorders

Aphasia, which is also expressed as language impairment or loss of language, occurs in 21–38% of stroke patients in the acute period. It is more common in ischemic stroke compared to haemorrhagic stroke. Aphasia negatively affects communication skills. Rehabilitation of aphasia should be an important goal of treatment in the acute and post-acute periods of stroke, bringing the patient into the community and workplace as soon as possible (Lazar & Boehme, 2017). Global aphasia is the most common aphasia in acute stroke. However, aphasia tends to functionally heal during the first year and always improves into a less severe form. Non-fluent aphasia can turn into fluent aphasia. Global aphasia may turn into Wernicke's aphasia. Broca's aphasia can turn into anomic aphasia (Pedersen, Vinter, & Olsen, 2004). In individuals who are right hand dominant, post-stroke aphasia is almost always associated with infarcts in the left hemisphere, but rarely (2–10%), a right hemisphere infarction can also cause aphasia. Aphasia is often associated with infarcts in areas irrigated by the MCA, such as the perisylvian cortex in the left hemisphere and the basal ganglia below, internal capsule and periventricular white matter (Berthier, 2005).

Early recognition of aphasia has positive effects on rehabilitation and neuronal recovery. The Minnesota Test, Porch index, Boston Aphasia diagnostic test, Token test, Western aphasia test, Frenchay aphasia screening test, Communication Ability in Daily Life, Schuell Test and acute aphasia screening protocol are examples of tests used to detect aphasia (Atamaz, 2007). Classical aphasia classification has been developed according to the effectiveness of speech fluency, comprehension, naming, expression and repetition. Global aphasia, Broca's aphasia and transcortical motor aphasia are examples of non-fluent aphasia, whereas Wernicke's aphasia, transcortical sensory aphasia and conduction aphasia are examples of fluent aphasia (de Oliveira, Marin, & Bertolucci, 2013). In motor speech disorders, such as dysarthria and speech apraxia, the deficit is in the motor production of speech.

Aphasia rehabilitation is applied by a multidisciplinary team and is arranged according to symptoms, localization of cerebral infarction, aetiology, speech level and cognitive skills. Since the results of the literature are contradictory, the time to start aphasia rehabilitation after stroke, how long the rehabilitation will last and its intensity have not been determined (Koyuncu & Özgirgin, 2016). Classic, pragmatic, behaviour modification, syndromic, Luria's functional, neurolinguistic, impairment-based or cognitive neuropsychological approach and social or consequences-based approach are the treatment options used in aphasia rehabilitation (Basso, Forbes, & Boller, 2013; Koyuncu & Özgirgin, 2016). The therapy can start with activities that will provide automatic speech, such as counting numbers and counting months. Afterwards, naming objects and pictures, choosing appropriate words and even more complicated activities, such as choosing sentences, can be studied. Textual studies, reading comprehension studies and attention studies contribute to recovery. Computer programs contribute to the development of the correct sentence structure through activities such as word finding and graphical clues (Koyuncu & Özgirgin, 2016). Examples of pharmacological agents that may be effective in the treatment of stroke-related aphasia are memantine (N-methyl-D-aspartate receptor antagonist) and donepezil (central dose-dependent acetylcholine esterase inhibitor) (Yang, 2017; Zhang et al., 2018). A review published in 2020 showed that TMS and tDCS can be effective in the treatment of post-stroke aphasia (Breining & Sebastian, 2020).

## Motor Disorders

Motor weakness is one of the most common stroke problems, occurring at rate of 80–90%. At least two-thirds of stroke patients present with hemiplegia. Unilateral motor paralysis of the body in a stroke patient is called hemiplegia. Among stroke patients, monoplegia occurs in 19%, paraplegia occurs in 1% and three or four limbs are affected in 2%. The level of motor weakness is important in stroke patients. Motor weakness has a significant effect on DLA. Failure to perform the movements as desired may be related to muscle weakness as well as neuropsychological function deficits. Neglect and apraxia negatively affect mobility, despite sufficient muscle strength. For this, evaluation of cognitive functions is recommended (Melo & Bogousslavsky, 2001).

Patients who start to develop synergy patterns after a stroke may not be able to use their muscles in isolation. It is not correct to evaluate the strength of a single muscle. Brunnstrom motor staging, which examines the flexor and extensor synergies and isolated muscle movements separated from synergy patterns, can be used to evaluate motor recovery after stroke. Control is as important as power in the formation of active movement (Karataş, 2011). The Fugl-Meyer Assessment and Motor Assessment Scale are other tests that evaluate motor function (Malouin, Pichard, Bonneau, Durand, & Corriveau, 1994). In the early period after a stroke, muscle tone decreases and later on, tonus increases, which may cause spasticity to occur. Spasticity can complicate movement, cause secondary complications and prolong rehabilitation. For this, muscle tone should be evaluated in follow-up visits (Karataş, 2011). The Modified Ashworth Scale is often used for the assessment of spasticity. Table 1 shows the Modified Ashworth Scale (Bohannon & Smith, 1987).

To ensure balance, intact vestibular, visual, proprioceptive, cognitive, cerebellar and motor function compatibilities are required. If a deficit develops in the sensorimotor system as a result of stroke, loss of balance occurs and an asymmetric distribution in body weight and inability to maintain posture develops. Improving posture, balance and ambulation in these patients should be one of the main goals of rehabilitation (Bittar & Barros, 2011).

*Table 1. Modified Ashworth Scale*

| Grade | Description |
|---|---|
| 0 | No increase in muscle tone |
| 1 | Slight increase in muscle tone, manifested by a catch and release, followed by minimal resistance at the end of the ROM when the affected part is moved in flexion or extension |
| 1+ | Slight increase in muscle tone, manifested by a catch, followed by minimal resistance throughout the remainder (less than half) of the ROM |
| 2 | More marked increase in muscle tone through most of the ROM, but the affected part is easily moved. |
| 3 | Considerable increase in muscle tone; passive movement is difficult |
| 4 | Affected part rigid in flexion or extension |

## Sensory Disorders

As a result of stroke, sensorial problems such as numbness, burning sensation, tingling, changes in temperature sensitivity, increased sensitivity to touch and pain and distribution of proprioception, stereognosis, tactile sensations and discrimination can be seen (Carey & Matyas, 2011; Carlsson, Gard, & Brogårdh, 2018; Connell, Lincoln, & Radford, 2008). These sensory deficits cause joint and skin damage due to deterioration in balance, coordination, motor control and protective sensation. Severe pain may occasionally occur in thalamic and spinothalamic tract lesions (Harvey et al., 2007).

## CONCLUSION

Stroke clinic forms according to the affected cerebral artery, the side of the lesion in the brain or the type of pathophysiology. Mental dysfunction, speech and language disorders, motor and sensory deficits can be seen in relation to these variables.

## REFERENCES

Al-Qazzaz, N. K., Ali, S. H., Ahmad, S. A., Islam, S., & Mohamad, K. (2014). Cognitive impairment and memory dysfunction after a stroke diagnosis: A post-stroke memory assessment. *Neuropsychiatric Disease and Treatment, 10*, 1677. doi:10.2147/NDT.S67184 PMID:25228808

Amarenco, P. (2001). Cerebellar stroke syndromes. In J. Bogousslavsky & L. R. Caplan (Eds.), *Stroke Syndromes* (pp. 540–556). Cambridge University Press. doi:10.1017/CBO9780511586521.043

Arboix, A., & Martí-Vilalta, J. L. (2009). Lacunar stroke. *Expert Review of Neurotherapeutics, 9*(2), 179-196. doi:10.1586/14737175.9.2.179

Atamaz, F. (2007). Communication disorders and their rehabilitation in stroke. *Turkish Journal of Physical Medicine and Rehabilitation, 53*(1), 11–15.

Azouvi, P., Jacquin-Courtois, S., & Luauté, J. (2017). Rehabilitation of unilateral neglect: Evidence-based medicine. *Annals of Physical and Rehabilitation Medicine, 60*(3), 191–197. doi:10.1016/j.rehab.2016.10.006 PMID:27986428

Basso, A., Forbes, M., & Boller, F. (2013). Rehabilitation of aphasia. In *Handbook of clinical neurology* (Vol. 110, pp. 325–334). Elsevier. doi:10.1016/B978-0-444-52901-5.00027-7

Berthier, M. L. (2005). Poststroke aphasia. *Drugs & Aging, 22*(2), 163–182. doi:10.2165/00002512-200522020-00006 PMID:15733022

Bittar, R. S. M., & Barros, C. G. C. (2011). Vestibular rehabilitation with biofeedback in patients with central imbalance. *Revista Brasileira de Otorrinolaringologia (English Ed.), 77*(3), 356–361. doi:10.1590/S1808-86942011000300014 PMID:21739011

Bohannon, R. W., & Smith, M. B. (1987). Interrater reliability of a modified Ashworth scale of muscle spasticity. *Physical Therapy, 67*(2), 206–207. doi:10.1093/ptj/67.2.206 PMID:3809245

Breining, B. L., & Sebastian, R. (2020). Neuromodulation in Post-stroke Aphasia Treatment. *Current Physical Medicine and Rehabilitation Reports*, *8*(2), 44–56. doi:10.100740141-020-00257-5 PMID:33344066

Buğdaycı, D., & Caf, N. (2016). Poststroke cognitive impairments, apraxia and neglect. *Turkiye Klinikleri J PM&R-Special Topics*, *9*(1), 17–23.

Carey, L. M., & Matyas, T. A. (2011). Frequency of discriminative sensory loss in the hand after stroke in a rehabilitation setting. *Journal of Rehabilitation Medicine*, *43*(3), 257–263. doi:10.2340/16501977-0662 PMID:21305243

Carlsson, H., Gard, G., & Brogårdh, C. (2018). Upper-limb sensory impairments after stroke: Self-reported experiences of daily life and rehabilitation. *Journal of Rehabilitation Medicine*, *50*(1), 45–51. doi:10.2340/16501977-2282 PMID:29068038

Çevikol, A., & Çakcı, A. (2015). İnme rehabilitasyonu [Stroke rehabilitation]. In H. Oğuz, H. Çakırbay, & B. Yanık (Eds.), *Tıbbi Rehabilitasyon* [Medical Rehabilitation] (pp. 419–448). Nobel Tıp Kitabevleri.

Chechlacz, M., Rotshtein, P., Bickerton, W.-L., Hansen, P. C., Deb, S., & Humphreys, G. W. (2010). Separating neural correlates of allocentric and egocentric neglect: Distinct cortical sites and common white matter disconnections. *Cognitive Neuropsychology*, *27*(3), 277–303. doi:10.1080/02643294.2010.519699 PMID:21058077

Connell, L. A., Lincoln, N., & Radford, K. (2008). Somatosensory impairment after stroke: Frequency of different deficits and their recovery. *Clinical Rehabilitation*, *22*(8), 758–767. doi:10.1177/0269215508090674 PMID:18678576

Crowfoot, G. M., van der Riet, P. J., & Maguire, J. M. (2016). Living within stories: Exploring the experiences of people with transient ischemic attack. *Nursing & Health Sciences*, *18*(1), 52–57. doi:10.1111/nhs.12227 PMID:26275055

Cumming, T. B., Marshall, R. S., & Lazar, R. M. (2013). Stroke, cognitive deficits, and rehabilitation: Still an incomplete picture. *International Journal of Stroke*, *8*(1), 38–45. doi:10.1111/j.1747-4949.2012.00972.x PMID:23280268

de Oliveira, F. F., Marin, S. M. C., & Bertolucci, P. H. F. (2013). Communicating with the non-dominant hemisphere: Implications for neurological rehabilitation. *Neural Regeneration Research*, *8*(13), 1236. doi:10.3969/j.issn.1673-5374.2013.13.009 PMID:25206418

Demeyere, N., & Gillebert, C. R. (2019). Ego-and allocentric visuospatial neglect: Dissociations, prevalence, and laterality in acute stroke. *Neuropsychology*, *33*(4), 490–498. doi:10.1037/neu0000527 PMID:30896236

Donnan, G. A., & Norrving, B. (2008). Lacunes and lacunar syndromes. *Handbook of Clinical Neurology, 93*, 559–575. doi:10.1016/S0072-9752(08)93027-X PMID:18804668

Donovan, N. J., Kendall, D. L., Heaton, S. C., Kwon, S., Velozo, C. A., & Duncan, P. W. (2008). Conceptualizing functional cognition in stroke. *Neurorehabilitation and Neural Repair, 22*(2), 122–135. doi:10.1177/1545968307306239 PMID:17761809

Durlanık, G. (2016). Vascular anatomy and correlation with the clinical features by stroke. *Turkiye Klinikleri J PM&R-Special Topics, 9*(1), 1–7.

Easton, J. D., Saver, J. L., Albers, G. W., Alberts, M. J., Chaturvedi, S., Feldmann, E., ... Kidwell, C. S. (2009). Definition and evaluation of transient ischemic attack: a scientific statement for healthcare professionals from the American Heart Association/ American Stroke Association Stroke Council; Council on Cardiovascular Surgery and Anesthesia; Council on Cardiovascular Radiology and Intervention; Council on Cardiovascular Nursing; and the Interdisciplinary Council on Peripheral Vascular Disease: the American Academy of Neurology affirms the value of this statement as an educational tool for neurologists. *Stroke, 40*(6), 2276–2293. doi:10.1161/ STROKEAHA.108.192218 PMID:19423857

Ferber, S., & Karnath, H.-O. (2001). How to assess spatial neglect-line bisection or cancellation tasks? *Journal of Clinical and Experimental Neuropsychology, 23*(5), 599–607. doi:10.1076/jcen.23.5.599.1243 PMID:11778637

Ferro, J. M. (2001). Hyperacute cognitive stroke syndromes. *Journal of Neurology, 248*(10), 841–849. doi:10.1007004150170067 PMID:11697519

Ferro, J. M. (2003). Cardioembolic stroke: An update. *Lancet Neurology, 2*(3), 177–188. doi:10.1016/S1474-4422(03)00324-7 PMID:12849239

Frizzell, J. P. (2005). Acute stroke: Pathophysiology, diagnosis, and treatment. *AACN Advanced Critical Care, 16*(4), 421–440. doi:10.1097/00044067-200510000-00002 PMID:16269890

Ganzer, C. A., Barnes, A., Uphold, C., & Jacobs, A. R. (2016). Transient ischemic attack and cognitive impairment: A review. *The Journal of Neuroscience Nursing, 48*(6), 322–327. doi:10.1097/JNN.0000000000000236 PMID:27824800

Harvey, R., Roth, E., & Yu, D. (2007). Rehabilitation in stroke syndromes. In R. L. Braddom (Ed.), *Physical Medicine & Rehabilitation* (pp. 1175–1112). Saunders Elsevier.

Hurford, R., Charidimou, A., Fox, Z., Cipolotti, L., & Werring, D. J. (2013). Domain-specific trends in cognitive impairment after acute ischaemic stroke. *Journal of Neurology*, *260*(1), 237–241. doi:10.100700415-012-6625-0 PMID:22865200

Karataş, G. (2011). İnme [Stroke]. In M. Beyazova & Y. G. Kutsal (Eds.), *Fiziksel Tıp ve Rehabilitasyon* [Physical Medicine and Rehabilitation] (pp. 2761–2783). Güneş Kitabevi.

Koyuncu, E., & Özgirgin, N. (2016). Neurogenic language and speech disorders and rehabilitation in stroke. *Turkiye Klinikleri J PM&R-Special Topics*, *9*(1), 8–16.

Lazar, R. M., & Boehme, A. K. (2017). Aphasia as a predictor of stroke outcome. *Current Neurology and Neuroscience Reports*, *17*(11), 83. doi:10.100711910-017-0797-z PMID:28929424

Malouin, F., Pichard, L., Bonneau, C., Durand, A., & Corriveau, D. (1994). Evaluating motor recovery early after stroke: Comparison of the Fugl-Meyer Assessment and the Motor Assessment Scale. *Archives of Physical Medicine and Rehabilitation*, *75*(11), 1206–1212. doi:10.1016/0003-9993(94)90006-X PMID:7979930

Markus, H. (2008). Stroke: Causes and clinical features. *Medicine*, *36*(11), 586–591. doi:10.1016/j.mpmed.2008.08.009

Mehndiratta, P., Smith, S. C., & Worrall, B. B. (2015). Etiologic stroke subtypes: updated definition and efficient workup strategies. *Current Treatment Options in Cardiovascular Medicine, 17*(1), 1-11. doi:10.1007/s11936-014-0357-7

Melo, T. P., & Bogousslavsky, J. (2001). Hemiparesis and other types of motor weakness. In J. Bogousslavsky & L. R. Caplan (Eds.), *Stroke Syndromes* (pp. 22–33). Cambridge University Press. doi:10.1017/CBO9780511586521.004

Murphy, S. J., & Werring, D. J. (2020). Stroke: Causes and clinical features. *Medicine*, *48*(9), 561–566. Advance online publication. doi:10.1016/j.mpmed.2020.06.002 PMID:32837228

Palsetia, D., Rao, G. P., Tiwari, S. C., Lodha, P., & De Sousa, A. (2018). The clock drawing test versus mini-mental status examination as a screening tool for dementia: A clinical comparison. *Indian Journal of Psychological Medicine*, *40*(1), 1–10. doi:10.4103/IJPSYM.IJPSYM_244_17 PMID:29403122

Pare, J. R., & Kahn, J. H. (2012). Basic neuroanatomy and stroke syndromes. *Emergency Medicine Clinics*, *30*(3), 601–615. doi:10.1016/j.emc.2012.05.004 PMID:22974640

Park, J. E. (2017). Apraxia: Review and update. *Journal of Clinical Neurology (Seoul, Korea), 13*(4), 317–324. doi:10.3988/jcn.2017.13.4.317 PMID:29057628

Pedersen, P. M., Vinter, K., & Olsen, T. S. (2004). Aphasia after stroke: Type, severity and prognosis. *Cerebrovascular Diseases (Basel, Switzerland), 17*(1), 35–43. doi:10.1159/000073896 PMID:14530636

Román, G. C. (2003). Vascular dementia: distinguishing characteristics, treatment, and prevention. *Journal of the American Geriatrics Society, 51*(5s2), S296-S304. doi:10.1046/j.1532-5415.5155.x

Sahathevan, R., Brodtmann, A., & Donnan, G. A. (2012). Dementia, stroke, and vascular risk factors; a review. *International Journal of Stroke, 7*(1), 61–73. doi:10.1111/j.1747-4949.2011.00731.x PMID:22188853

Salter, K., Teasell, R., Foley, N., & Allen, L. (2013). *Evidence-Based Review of Stroke Rehabilitation: Cognitive Disorders and Apraxia*. Academic Press.

Schoen, J. C., Boysen, M. M., Warren, C. R., Chakravarthy, B., & Lotfipour, S. (2011). Vertebrobasilar artery occlusion. *The Western Journal of Emergency Medicine, 12*(2), 233. PMID:21691534

Stone, S., Wilson, B., Wroot, A., Halligan, P., Lange, L., Marshall, J., & Greenwood, R. (1991). The assessment of visuo-spatial neglect after acute stroke. *Journal of Neurology, Neurosurgery, and Psychiatry, 54*(4), 345–350. doi:10.1136/jnnp.54.4.345 PMID:2056321

Sun, J.-H., Tan, L., & Yu, J.-T. (2014). Post-stroke cognitive impairment: Epidemiology, mechanisms and management. *Annals of Translational Medicine, 2*(8). Advance online publication. doi:10.3978/j.issn.2305-5839.2014.08.05 PMID:25333055

Ukraintseva, S., Sloan, F., Arbeev, K., & Yashin, A. (2006). Increasing rates of dementia at time of declining mortality from stroke. *Stroke, 37*(5), 1155–1159. doi:10.1161/01.STR.0000217971.88034.e9 PMID:16601210

Yang, S. N. (2017). Current Evidence for Post Stroke Aphasia Treatment. *Brain & Neurorehabilitation, 10*(2), e15. Advance online publication. doi:10.12786/bn.2017.10.e15

Zhang, X., Shu, B., Zhang, D., Huang, L., Fu, Q., & Du, G. (2018). The efficacy and safety of pharmacological treatments for post-stroke aphasia. *CNS & Neurological Disorders-Drug Targets, 17*(7), 509-521. doi:10.2174/1871527317666180706143051

# Chapter 9
# Complications of Stroke

**Alper Uysal**
*Hatay Training and Research Hospital, Turkey*

## ABSTRACT

*Stroke is one of the leading causes of disability and mortality and can cause a serious socioeconomic burden. Some of the comorbidities and secondary complications of stroke can threaten the patient's life or cause serious pain or negatively affect the patient's involvement in rehabilitation or worsen daily life activities or make it difficult to bring the patient into the community and workplace. This chapter focuses on the symptoms and signs, diagnosis, and management of these comorbidities and complications. It highlights diagnosis and treatment of cardiac problems, sleep disorders, deep vein thrombosis, pulmonary embolism, dysphagia, malnutrition, and pneumonia. Depression, central post-stroke pain, upper limb problems after stroke, spasticity, bladder dysfunction are also discussed in this chapter.*

## INTRODUCTION

Loss of physical function and disability may continue after the acute period in people who have had a stroke. Every year, worldwide, one-third of the people who have a stroke remain severely disabled. As time progresses, other medical, musculoskeletal and psychosocial disorders may be added to stroke-related neurological deficits (Chohan, Venkatesh, & How, 2019). These medical complications prolong the hospital stay of the patients and increase the treatment costs. They can even cause death in the acute and subacute periods. They may adversely affect functional recovery and cause inadequate recovery. They can negatively affect the rehabilitation process (Kumar, Selim, & Caplan, 2010). Stroke not only significantly reduces the

DOI: 10.4018/978-1-7998-7433-1.ch009

patient's quality of life, but also increases the financial and psychological burden of the patient and his family.

Most complications develop in the first weeks after a stroke. While some complications such as heart problems, dysphagia and pneumonia occur in the first days after stroke, venous thrombosis, bedsores and falls may occur later (Indredavik, Rohweder, Naalsund, & Lydersen, 2008; Johnston et al., 1998; Kumar et al., 2010). This article sheds light on the clinical manifestations, diagnosis and treatment of stroke complications.

## CARDIAC PROBLEMS

The second most common cause of deaths after stroke is cardiac diseases. Cardiac problems cause stroke, but they can also occur as a result of stroke (Chen et al., 2017). Cardiac problems, such as myocardial infarction, congestive heart failure and atrial fibrillation, can be seen in the first 3 days after stroke (Scheitz, Nolte, Doehner, Hachinski, & Endres, 2018). Common risk factors for heart disease and stroke are age, hypertension, hyperlipidaemia and diabetes mellitus (Chen et al., 2017). The control of the sympathovagal outflow to the heart is disturbed as a result of infarction in the brain tissues. There is strong evidence that this situation can be effective in the pathophysiology of post-stroke cardiac problems. As a result of excessive stimulation of the autonomic nervous system and the hypothalamic-pituitary-adrenal axis, excessive levels of cortisone and catecholamine are secreted. Thus, calcium flows into the cytosol in the heart muscle and contraction occurs, making relaxation difficult. Direct myocardial catecholamine toxicity, microvascular dysfunction and coronary demand ischemia are mechanisms that explain the pathophysiology of stroke hearth syndrome (Scheitz et al., 2018).

Stroke, one of the leading causes of long-term disability, causes a decrease in patients' daily life activities (DLA) and an increased risk of falling. In general, patients with a history of stroke use more oxygen than healthy people during walking. People with stroke need more energy when doing daily household chores, such as gathering, wiping and sweeping the home. The increased energy and oxygen demands also create an additional burden for the heart. Activities that require more energy can exacerbate mild cardiac problems.

Stroke patients should be monitored by electrocardiogram (ECG) before an exercise program and evaluated with a gradual exercise program. The change of heart rate and rhythm and systolic and diastolic blood pressures should be examined throughout exercise. Rehabilitation activities should be arranged according to the severity of cardiac problems and the exercise capacity of the person. The most appropriate rehabilitation program should include personalized exercises. Exercise programs with treadmills or bicycle ergometers should be prescribed 20–60 minutes a day, 3–7 days a week with an intensity of 40–70% of maximum oxygen consumption or heart rate reserve (Gordon et al., 2004). In order to prevent cardiovascular problems that may develop during post-stroke rehabilitation, criteria for stopping rehabilitation have been developed by Fletcher et al. (Table 1) (Fletcher, Dunbar, Coleman, Jann, & Fletcher, 1993).

*Table 1. Termination criteria of rehabilitation for cardiac reasons*

| 1. Beginning cardiopulmonary symptoms |
|---|
| 2. Pulse reduction of more than 20% |
| 3. More than 50% increase in heart rate |
| 4. Systolic blood pressure rising to 240 mmHg and above |
| 5. Systolic blood pressure decreasing by 30 mmHg or more or falling below 90 mmHg |
| 6. Diastolic blood pressure rising to 120 mmHg and above |
| A positive one is enough to end the activity. |

## SLEEP DISORDERS

Approximately 60–70% of stroke patients experience sleep disturbance breathing (SDB). The most common form of SDB in stroke patients is obstructive sleep apnoea (OSA). Patients with sleep disorders have difficulty falling asleep at night. Symptoms such as snoring, stridor, apnoea, irregular respiration, shortness of breath and sudden awakenings are observed during sleep. Due to poor quality and insufficient sleep, these patients experience headaches, fatigue, irritability, depression, excessive sleepiness, and attention and memory problems during the day. As a result, the participation of the patient in rehabilitation is reduced and functional activity and quality of life are also negatively affected.

The aetiology of post-stroke sleep disorder respiration includes brain infarction, secondary complications associated with brain infarction (aspiration, pulmonary oedema, inactivity, pneumonia and autonomic changes) and diseases that also existed before stroke, such as OSA and heart failure. As a result of SDB, especially OSA, brain blood flow and sympathetic activity are increased, endothelial function is impaired and platelet aggregation is increased. Consequently, the risk of stroke also increases with increasing cardiovascular risk (Bassetti, 2005). Symptoms of SDB, such as excessive daytime sleepiness and prolonged sleep, may be independent risk factors for stroke or may also be the result of stroke. In the diagnosis of SDB, respiratory polygraphy, oximetry and polysomnography tests can be used.

In the treatment of post-stroke sleep disorder, secondary complications should be prevented and treated early if they occur. Caution should be exercised in the use of drugs that suppress respiration. Weight loss and a lateral lying position should be recommended. Nasal continuous positive airway pressure can be used as a treatment option for OSA (Bassetti & Hermann, 2011).

## DEEP VEIN THROMBOSIS AND PULMONARY EMBOLISM

Deep vein thrombosis (DVT) is a common pathology in the aetiology of pulmonary embolism (PE). The incidence of DVT in immobile patients varies between 10% and 75% according to the diagnostic method used. Most DVTs are asymptomatic. Symptomatic DVT prevalence is 2–10%. DVT develops most frequently on post-stroke days 7 and 8. If treatment is not given, 15% of patients die. Inactivity, flask tonus of the lower extremity, atrial fibrillation and tissue plasminogen activator (TPA) treatment increases the risk of DVT. The frequency of development of venous thromboembolism (VTE) has increased in patients with subarachnoid haemorrhage (SAH) or intracerebral haemorrhage (ICH).

The first choice test in the diagnosis of DVT is ultrasound; Doppler techniques also support the diagnosis. Other tests that can be used to diagnose DVTs are venography, magnetic resonance imaging (MRI) and computed tomography (CT). In DVT prophylaxis, unfractionated heparin and low molecular weight heparin can be used, but caution should be exercised, as the risk of bleeding increases slightly. Pneumatic sequential compression devices reduce the risk of DVT (Khan et al., 2017). Anti-embolism stockings should not be prescribed to patients with acute stroke for DVT prophylaxis. In immobile stroke patients, pneumatic compression devices can be used for DVT prophylaxis in the acute period. If a pneumatic compression device is to be used, it should be started within the first 3 days after the stroke. It should be used for the first month or until the patient is discharged or mobilized (National Institute for Health Care Excellence, 2018).

DVT can be seen especially in immobile patients during the rehabilitation phase and if left untreated, it may cause PE. Therefore, care should be taken. In prospective studies evaluating venous thromboembolic complications, the incidence of PE was found to be 10–13%. Asymptomatic proximal DVT can cause fatal PE. PE originating from a distal DVT (distal to the knee) is smaller, clinically quieter and less lethal than PE originating from a proximal DVT. However, 20% of untreated distal DVTs spread proximally. In the DVT clinic, swelling, pain and venous ulcer can be seen in the lower extremity (Kelly, Rudd, Lewis, & Hunt, 2001). In the case of developing PE, symptoms and findings such as dyspnoea, pleuritic chest pain, cough, haemoptysis, syncope, hypotension, tachycardia, tachypnoea and hypoxemia may be seen in the patient (Iurato et al., 2015). Arterial blood gases, chest X-ray, ECG, ultrasonography (USG) (lower extremity), venography and pulmonary CT angiography are diagnostic tests for PE (Caplan & Kumar, 2016).

In the treatment of PE, although anticoagulant therapy (low molecular weight heparin, unfractionated heparin, fondaparinux, factor Xa inhibitors and warfarin) is sufficient in hemodynamically stable patients, in hemodynamically unstable patients, thrombolytic therapy or pulmonary embolectomy is performed as a first choice and then anticoagulant agents can be applied. Fibrinolytic agents are used in the treatment of massive PEs. Long-term anticoagulant therapy is important in preventing relapse. Vena cava filters can be used in cases where anticoagulant therapy is contraindicated and in patients with recurrent PE despite anticoagulant therapy (Iurato et al., 2015).

## DYSPHAGIA, MALNUTRITION AND PNEUMONIA

Dysphagia is difficulty swallowing. The incidence of dysphagia is 22–65% in the acute period, but 2–4 weeks post-stroke, more than 80% of patients recover their ability to swallow. The presence of dysphagia is a sign of poor prognosis in stroke (D. J. Ramsey, Smithard, & Kalra, 2003). Dysphagia facilitates the development of malnutrition, dehydration, aspiration and pulmonary infections (Smithard et al., 1996). It is more common in brainstem and medullary strokes. Dysphagia can cause aspiration. Terre et al. found that aspiration after stroke occured in 2 out of every 3 patients and half of this was silent aspiration. The patient has complaints such as cough during feeding, change in voice after feeding and difficulty swallowing. Aspiration is more common in brainstem and large vessel strokes. Aspiration should be suspected in patients with cough, dysphagia and pathological signs on direct chest radiography (Terre & Mearin, 2006).

Daniels et al. showed that silent aspiration developed in 25% of patients within the first 5 days after stroke (Daniels et al., 1998). In silent aspiration, the patient does not have any cough or any signs of distress, but aspiration can be demonstrated on videofluoroscopy (VF) and fibreoptic endoscopic evaluation of swallowing (FEES) (D. Ramsey, Smithard, & Kalra, 2005). If the pulse oximeter is combined with a bedside swallow examination, the sensitivity of the tests increases (Smith, Lee, O'Neill, & Connolly, 2000). The swallowing skills of the patients should be evaluated with bedside examination methods. In the water swallowing test (WST), which is a screening test, 3–150 ml of water was used in the literature. Due to the risk of aspiration, the modified WST using 3 ml of water has been developed. But, there are cases when it is recommended to use 30 ml of water for the WST, since the swallowing reflex may be difficult if the oral cavity is not wide enough (Osawa,

Maeshima, & Tanahashi, 2013). VF is the gold standard test for verification of aspiration (Mari et al., 1997).

If the patient still cannot be fed on the 5th day post-stroke, feeding with a nasogastric feeding tube can be provided. A nasogastric feeding tube is thought not to prevent aspiration and therefore pneumonia. Nutrition can also be provided with a percutaneous endoscopic gastrostomy (PEG) tube, but a PEG tube should not be used in the early period, as it may result in poor results if preferred in the early period. Nearly 25% or more of patients with a PEG tube will recover at a level that does not require its long-term use (Caplan & Kumar, 2016). Although a nasogastric feeding tube has been used for 3–4 weeks, if there is still not enough improvement, a PEG feeding tube should be started.

The patient's diet should include mash, blendered, ground or chopped food and thickened liquids (Pretorius & Van Zyl, 2008). Fluid, protein and total caloric intake of patients with stroke should be carefully monitored during the rehabilitation process. Oral nutritional supplements may be beneficial; it should be kept in mind that a deficiency of micronutrients (folic acid, B complex and D vitamins, antioxidants and zinc) may cause cerebrovascular changes and increase the risk of stroke (Bouziana & Tziomalos, 2011).

Postural adjustments of the patient, swallow manoeuvres and rehabilitation are other methods to be used in treatment. The purpose of postural adjustments is to provide safe eating and drinking without aspiration. Examples of postural adjustments are body posture changes, such as side lying and lying down and head posture changes, such as head extension (chin up), head flexion (chin tuck) and head rotation (head turn). Swallow manoeuvres should be selected according to the type of swallowing disorder. Supraglottic swallow, super supraglottic swallow, effortful swallow and the Mendelsohn manoeuvre are the major swallowing manoeuvres used to treat dysphagia. Swallow rehabilitation improves swallowing function with an exercise-oriented approach and prevents complications, such as aspiration and pneumonia. Examples of exercise-based swallow rehabilitation are lingual resistance, Shaker head-lift, expiratory muscle strength training and the McNeill dysphagia therapy program. Pneumonia is seen in nearly 1 out of every 3 stroke patients in the acute period. Pneumonia is associated with about 35% of deaths after stroke. In the post-stroke patient, pneumonia may develop due to a swallowing deficit and aspiration (Sura, Madhavan, Carnaby, & Crary, 2012).

## DEPRESSION

Post-stroke depression (PSD) is among the most common emotional disorders in stroke patients. In a meta-analysis, the prevalence of PSD was found to be 29%

(Ayerbe, Ayis, Wolfe, & Rudd, 2013). In a literature review, the prevalence of major depression in rehabilitation clinics in the acute period was 21.7% and the prevalence of minor depression was 19.5%. In outpatients, the prevalence of major depression was 24% in the period from 3 months to 3 or more years after stroke, while the prevalence of minor depression was 23.9% (Robinson & Spalletta, 2010).

When compared to each other, stroke or orthopaedic trauma patients with equivalent physical disorders, depression was found to be more common in stroke patients. The severity of PSD has been shown to be associated with stroke severity, DLA score, social functioning and degree of cognitive function distortion. Patients with PSD who responded favourably to nortriptyline or fluoxetine treatment were found to have a better level of improvement in DLA. Mortality was higher in patients with PSD than stroke patients without depression.

The relationship between depression and lesion localization has been investigated and there are studies showing that there is a relationship between left frontal lobe infarction and the development of PSD, despite conflicting literature results (Robinson & Jorge, 2016). The presence of PSD causes an increase in disability and poor rehabilitation results (Paolucci et al., 2001). The risk of PSD is high in patients with pre-stroke psychiatric disease, under 70 years of age, high levels of neuroticism, positive family history, high neurological deficits and disability and women (Shi, Yang, Zeng, & Wu, 2017). Depression and stroke are interrelated, both being risk factors for each other. PSD negatively affects patient participation in the rehabilitation program and reduces the effectiveness of rehabilitation.

It is important to recognize and treat PSD early. Diagnosis of PSD can be accomplished with the Diagnostic and Statistical Manual of Mental Disorders (DSM-V) during the patient interview. Other tests that can help with the diagnosis are Hamilton Rating Scale for Depression, Beck Depression Inventory, Montgomery-Asberg Depression Rating Scale, Center for Epidemiological Studies Depression Scale and the Post-Stroke Depression Rating Scale. The most preferred antidepressants in treatment were the selective serotonin reuptake inhibitors (SSRIs), but SSRIs can cause intracranial bleeding. Citalopram, escitalopram, fluoxetine, fluvoxamine, paroxetine and sertraline are examples of SSRIs. Tricyclic antidepressants (TCAs) have more side effects in patients with stroke compared to SSRIs (Villa, Ferrari, & Moretti, 2018). In patients who develop PSD, the recovery rate of depression is in the range of 15–57% 1 year post-stroke (Ayerbe et al., 2013).

## CENTRAL POST-STROKE PAIN

Post-stroke pain has many causes, including central pain, hemiplegic shoulder pain, spasticity-based pain, headache and articular pain (Harrison & Field, 2015). The prevalence of pain in stroke patients varies between 19% and 74% depending on study design. The prevalence of central post-stroke pain (CPSP) was found to be 8–35% in the literature (Jönsson, Lindgren, Hallström, Norrving, & Lindgren, 2006). CPSP is a central neuropathic pain that develops as a result of central nervous system lesions. Patients with CPSP have complaints such as burning, freezing, squeezing, cold and prickling sensations. Sensory deficits such as hypoesthesia, hyperesthesia, hyperpathia, allodynia, paraesthesia, dysesthesia and hyperalgesia can be seen in the half of the body on the contralateral side of the cerebral lesion in patients with CPSP.

CPSP negatively affects the patient's sleep, quality of life and participation in rehabilitation and creates a predisposition to depression and anxiety (Harrison & Field, 2015). In the medical treatment of CPSP, antidepressants (amitriptyline and SSRIs), anticonvulsants (lamotrigine, gabapentin and pregabalin) and opioids are used. For treatment of drug-resistant cases, repetitive transcranial magnetic stimulation, motor cortex stimulation and deep brain stimulation can be used (Delpont et al., 2018; Lewis & Kriukelyte, 2019).

## POST-STROKE SHOULDER PAIN

In all, 25–50% of patients experience post-stroke shoulder pain (Delpont et al., 2018). Shoulder subluxation, hyperspastic hemiplegia, periarthritis of the shoulder, rotator cuff problems, bicipital tendinitis, soft tissue trauma, shoulder-hand syndrome, CPSP and avulsion injuries of the brachial plexus can cause shoulder pain in stroke patients. The cause of shoulder pain in stroke patients is often multifactorial. Incorrect position and handling of the hemiplegic arm can increase shoulder pain. Improper transfer of the patient or incomplete handling of the hemiplegic arm causes trauma by increasing the load on the shoulder joint. In the presence of sensory and perception deficits, protection against this trauma is reduced, so the pain may increase. It has been found that shoulder pain is more common in patients who need help during transfer (Koog, Jin, Yoon, & Min, 2010; Walsh, 2001). Electrical stimulation application has been found to improve the shoulder subluxation and limitation of passive shoulder lateral rotation (Koog et al., 2010).

### Shoulder Subluxation

Shoulder subluxation is a common cause of shoulder pain in stroke patients. Subluxation reduces the patient's participation in rehabilitation. The most common type of shoulder subluxation in stroke patients is inferior subluxation. In the post-stroke flask phase, the hemiplegic arm is pulled down with the effect of gravity. In healthy individuals, the supraspinatus and deltoid muscles contract and form opposing forces. Thus, joint integrity is preserved. However, in stroke patients, the decrease or loss of strength in the supraspinatus and deltoid muscles allows the humerus to be pulled down. The applied stretching forces and periarticular soft tissue injuries also contribute to the formation of subluxation.

However, muscle, ligament and capsule shortening and tonus increase may cause the development of subluxation in the spastic period (Vafadar, Côté, & Archambault, 2015). The origin of shoulder problems in stroke patients is the disruption of shoulder biomechanics. Pathological movement patterns emerge as a result of stroke. The anatomical alignment of the joints and the length and strength directions of the shoulder girdle muscles are affected and mechanical problems develop. Mobility of the scapula on the thorax, scapulohumeral movement rhythm and the stability of the humerus are impaired. Subluxation and shoulder pain occur as a result of these changes (Andersen, 1985).

The distance between the head of the humerus and acromion can be evaluated by palpation during physical examination. Some researchers evaluate the level of subluxation by determining how many fingers fit into this space. Assessment can also be made by measuring the distance between the acromion and the head or the lateral epicondyle of the humerus with a tape measure.

In the treatment, especially in the flaccid period, the arm is supported, properly positioned and care is taken not to damage the arm while helping with the transfer. A few authors have suggested that a pillow can be used to support the affected arm of the bedridden patients. From the moment the patient begins to sit, forearm support and an arm board can be used while in a wheelchair. Although shoulder straps are recommended for arm support, their use is still controversial. Some researchers argue that shoulder straps increase flexor synergy and spasticity by keeping the arm in the flexion position and may cause shoulder hand syndrome. They also argue that the use of sling creates a balance problem by disrupting the arm component of walking (Paci, Nannetti, & Rinaldi, 2005).

Another treatment option in the treatment of shoulder subluxation is functional electrical stimulation (FES). Chantraine et al. found a statistically significant increase in the improvement of functional level, subluxation and pain in the FES group compared to the control group (Chantraine, Baribeault, Uebelhart, & Gremion, 1999). According to a meta analysis published in 2015, FES reduces hemiplegic shoulder subluxation in the acute phase. However, no improvement in pain or upper limb functions or in the chronic period was shown (Vafadar et al., 2015).

## Adhesive Capsulitis

Adhesive capsulitis is another musculoskeletal disorder that causes shoulder pain. Shoulder pain causes stroke patients to avoid using his/her arm. Atrophy, contracture and immobilization facilitate the development of adhesive capsulitis. Adhesive changes are thought to occur in the capsule as a result of prolonged movement restriction, chronic tissue damage and inflammation. Joint range of motion (ROM), especially external rotation, is significantly reduced (McKenna, 2001).

The diagnosis can be made by arthrography and physical examination. It should be distinguished whether the decrease in ROM is of a spastic or contracture origin. Swath-type slings should not be used. Performing daily ROM exercises is important in the prevention of adhesive capsules. The effectiveness of intra-articular injections in the treatment of adhesive capsulitis is controversial (Kendall, 2010).

## Spasticity

Van Ouwenaller et al. found that the incidence of shoulder pain in spastic patients was higher than in flask patients (Van Ouwenaller, Laplace, & Chantraine, 1986). Hecht et al. showed the relationship of subscapularis muscle spasticity with ROM limitation and pain (Hecht, 1992). If a patient has spasticity in the subscapularis and pectoralis major muscles, which are the internal rotators of the arm, pain and passive ROM limitation can be managed by intramuscular injections of botulinum toxin type A (Teasell, Foley, Pereira, Sequeira, & Miller, 2012).

## Rotator Cuff Problems

To prevent impingement in healthy individuals, internal rotation of the humerus occurs during flexion of the arm and external rotation of the humerus occurs during the abduction of the arm. While the main muscle responsible for internal rotation is the subscapularis, the main muscle that performs external rotation is the infraspinatus. Any deficit in this mechanism causes impingement in the long head of the biceps or supraspinatus by bringing the tuberculum majus closer to the acromion during elevation of the upper limb. Spasticity or muscle weakness can also disrupt this mechanism and cause rotator cuff syndrome. In rotator cuff weakness, the humerus head is displaced upwards and hits the acromion, causing the supraspinatus to undergo impingement (Turner-Stokes & Jackson, 2002).

## Brachial Plexus Lesion

Muscle tone, tendon reflexes, sensory deficits, distribution of muscle weakness, pathological reflexes, muscle atrophy and fasciculation reflexes should be evaluated in stroke patients. An electromyography (EMG) study should be performed in the presence of findings that do not suggest a first motor neuron lesion. It should be kept in mind that, especially in patients with shoulder subluxation, a brachial plexus lesion may develop with the traction mechanism (Yu, 2004).

### Shoulder Hand Syndrome

Generally, shoulder pain occurs first, then pain and oedema develop in the hand and wrist. Pain and swelling in the hand and wrist are observed in all cases, while shoulder involvement occurs in half of the cases. Wrist, metacarpophalangeal and proximal interphalangeal joints are painful and passive flexion of these joints is limited due to oedema in the dorsum of the hand. The origin of this hand oedema is not lymphedema. Pain, tenderness and decreased ROM can be seen in the shoulder (Geurts, Visschers, Van Limbeek, & Ribbers, 2000). Shoulder hand syndrome usually occurs in the first 3 months after stroke.

Hyperesthesia, vasomotor disorders and trophic changes in the upper limb muscle and skin are other symptoms and signs that may accompany the clinical picture. In these patients, patch demineralization in the bone can be demonstrated with X-ray. Bone scintigraphy is more valuable in diagnosis (Turner-Stokes & Jackson, 2002). Treatment should be started early. TCAs (imipramine and amitriptyline), anticonvulsants (gabapentin and carbamazepine), nonsteroidal anti-inflammatory drugs (NSAIDs), corticosteroids, free-radical scavengers (vitamin C), opioids and bisphosphonates are agents that can be used in pharmacological treatment (Harden, 2005). In physical therapy, contrast bath, desensitization, ROM exercises and strengthening exercises can be applied. Stellate ganglion block or surgical sympathectomy are other treatment options (Harden et al., 2013).

## SPASTICITY

Approximately 25% of patients with stroke have spasticity (Sommerfeld, Gripenstedt, & Welmer, 2012). Spasticity is a component of first motor neuron disease. It negatively affects perineal care, rehabilitation, pain, ambulation and DLA in stroke patients. Risk factors for the development of spasticity are severe paresis, brain stem infarction, haemorrhagic infarction and young age. Spasticity does not always have to be treated, as it also has some positive effects. For example, quadriceps femoris spasticity helps patients stand. Spasticity prevents the decrease in the amount of muscle and slows the development of osteoporosis. Untreated spasticity can cause secondary complications, such as contractures (Francisco & McGuire, 2012).

While non-pharmacological treatment options for spasticity are stretching exercises, thermotherapy, ultrasound, electrical stimulation, biofeedback, hydrotherapy, robotics and orthoses, pharmacological treatment options are local (alcohol, phenol and botulinum toxin) and systemic (baclofen, benzodiazepines, tizanidine and dantrolene) drugs. Intrathecal baclofen pump, tendon lengthening surgeries, tendon transfers, neurolysis and selective dorsal rhizotomy are interventional procedures applied in the treatment of spasticity (Francisco & McGuire, 2012; Thibaut et al., 2013).

## BLADDER DYSFUNCTION

More than 50% of patients develop bladder dysfunction in the first week after a stroke. However, in the first year, bladder dysfunction improves in many patients (Zorowitz & Tietjen, 1999). Many studies have shown an association between the presence of urinary incontinence and mortality or poor functional recovery in stroke patients. If the neuromicturition pathways are affected due to infarction, a hyperactive bladder develops. This type of bladder disorder is the most common type in stroke patients and is characterized by urges to urinate before the bladder is full, but the bladder cannot be emptied adequately after urination.

In the treatment of hyperactive bladder, a patient-specific bladder retraining program should be organized and a timed voiding program should be implemented for at least 6 weeks. After stroke, the cortical awareness of bladder filling decreases, so in the timed voiding program, voiding is performed every 2–4 hours regardless of the patient's need. Timed voiding intervals can be opened according to the level of incontinence recovery. Clean intermittent catheterization can be performed for incontinence and anticholinergic drugs can be used for the treatment of hyperactive bladder.

In the acute period after a stroke, the flask bladder may develop and its prevalence ranges from 21% to 35%. However, diabetic polyneuropathy or anticholinergic drugs can cause flask bladder during chronic stroke. In this type of bladder dysfunction, detrusor contraction is insufficient. Since the bladder is full, the urine flows as leakage. Insufficient cognition in stroke patients can also cause incontinence (Mehdi, Birns, & Bhalla, 2013; Ostaszkiewicz, Johnston, & Roe, 2004). The residual urine volume of the patient should be measured by catheterization after voiding. If medical treatment is to be started for incontinence, urodynamic evaluation should be performed first. Catheterization can be used in the treatment of hypoactive detrusor. A permanent catheter should not be used as much as possible since it may cause infection (Zorowitz & Tietjen, 1999). In case of residual urine, clean intermittent catheterization can be performed. Presence of residual urine over 100 cc is a risk factor for infection. Clean intermittent catheterization should be performed every 4–6 hours. If voluntary voiding is present, the patient should voluntarily urinate before the application. The intervals of intermittent catheterization may open after spontaneous voiding begins and may be terminated when less than 100 ml of urine remains after voiding.

In the presence of hyperactive detrusor, anticholinergic agents, such as oxybutynin and tolterodine, can be used (Di Benedetto, 2011; Dorsher & McIntosh, 2012). In the treatment of bladder dysfunction, pelvic floor exercises, electrical stimulation and biofeedback may be useful treatment choices (Cournan, 2012; Nelson, 2007; Shen & Liu, 2018).

## CONCLUSION

As a result, various medical, musculoskeletal and psychosocial complications may develop over time in stroke patients. These complications can add to the neurologic disability associated with stroke, further reducing the quality of life or reducing participation in rehabilitation or prolonging the hospital stay. By knowing enough about these complications, precautions can be taken before many complications occur or if they develop, they can be detected and treated in the early period.

## REFERENCES

Andersen, L. T. (1985). Shoulder pain in hemiplegia. *The American Journal of Occupational Therapy*, *39*(1), 11–19. doi:10.5014/ajot.39.1.11 PMID:2579561

Ayerbe, L., Ayis, S., Wolfe, C. D., & Rudd, A. G. (2013). Natural history, predictors and outcomes of depression after stroke: Systematic review and meta-analysis. *The British Journal of Psychiatry*, *202*(1), 14–21. doi:10.1192/bjp.bp.111.107664 PMID:23284148

Bassetti, C. L. (2005). Sleep and stroke. *Seminars in Neurology*, *25*(1), 19–32. doi:10.1055-2005-867073 PMID:15798934

Bassetti, C. L., & Hermann, D. M. (2011). Sleep and stroke. In *Handbook of clinical neurology* (Vol. 99, pp. 1051–1072). Elsevier. doi:10.1016/B978-0-444-52007-4.00021-7

Bouziana, S. D., & Tziomalos, K. (2011). Malnutrition in patients with acute stroke. *Journal of Nutrition and Metabolism*, *2011*, 1–7. doi:10.1155/2011/167898 PMID:22254136

Caplan, L. R., & Kumar, S. (2016). Complications in stroke patients. In L. R. Caplan (Ed.), Caplan's Stroke: A Clinical Approach (5th ed., pp. 594-607). Cambridge University Press. doi:10.1017/CBO9781316095805.020

Chantraine, A., Baribeault, A., Uebelhart, D., & Gremion, G. (1999). Shoulder pain and dysfunction in hemiplegia: Effects of functional electrical stimulation. *Archives of Physical Medicine and Rehabilitation*, *80*(3), 328–331. doi:10.1016/S0003-9993(99)90146-6 PMID:10084443

Chen, Z., Venkat, P., Seyfried, D., Chopp, M., Yan, T., & Chen, J. (2017). Brain–heart interaction: Cardiac complications after stroke. *Circulation Research*, *121*(4), 451–468. doi:10.1161/CIRCRESAHA.117.311170 PMID:28775014

Chohan, S. A., Venkatesh, P. K., & How, C. H. (2019). Long-term complications of stroke and secondary prevention: An overview for primary care physicians. *Singapore Medical Journal*, *60*(12), 616–620. doi:10.11622medj.2019158 PMID:31889205

Cournan, M. (2012). Bladder management in female stroke survivors: Translating research into practice. *Rehabilitation Nursing*, *37*(5), 220–230. doi:10.1002/rnj.054 PMID:22949275

Daniels, S. K., Brailey, K., Priestly, D. H., Herrington, L. R., Weisberg, L. A., & Foundas, A. L. (1998). Aspiration in patients with acute stroke. *Archives of Physical Medicine and Rehabilitation*, *79*(1), 14–19. doi:10.1016/S0003-9993(98)90200-3 PMID:9440410

Delpont, B., Blanc, C., Osseby, G., Hervieu-Bègue, M., Giroud, M., & Béjot, Y. (2018). Pain after stroke: A review. *Revue Neurologique*, *174*(10), 671–674. doi:10.1016/j.neurol.2017.11.011 PMID:30054011

Di Benedetto, P. (2011). Clean intermittent self-catheterization in neuro-urology. *European Journal of Physical and Rehabilitation Medicine*, *47*(4), 651–659. PMID:22222962

Dorsher, P. T., & McIntosh, P. M. (2012). Neurogenic bladder. *Advances in Urology*, *816274*, 1–16. Advance online publication. doi:10.1155/2012/816274 PMID:22400020

Fletcher, B. J., Dunbar, S., Coleman, J., Jann, B., & Fletcher, G. F. (1993). Cardiac precautions for non-acute inpatient settings. *American Journal of Physical Medicine & Rehabilitation*, *72*(3), 140–143. doi:10.1097/00002060-199306000-00006 PMID:8512675

Francisco, G. E., & McGuire, J. R. (2012). Poststroke spasticity management. *Stroke, 43*(11), 3132–3136. doi:10.1161/STROKEAHA.111.639831 PMID:22984012

Geurts, A., Visschers, B., Van Limbeek, J., & Ribbers, G. (2000). Systematic review of aetiology and treatment of post-stroke hand oedema and shoulder-hand syndrome. *Scandinavian Journal of Rehabilitation Medicine, 32*(1), 4–10. doi:10.1080/003655000750045668 PMID:10782934

Gordon, N. F., Gulanick, M., Costa, F., Fletcher, G., Franklin, B. A., Roth, E. J., & Shephard, T. (2004). Physical activity and exercise recommendations for stroke survivors: An American Heart Association scientific statement from the Council on Clinical Cardiology, Subcommittee on Exercise, Cardiac Rehabilitation, and Prevention; the Council on Cardiovascular Nursing; the Council on Nutrition, Physical Activity, and Metabolism; and the Stroke Council. *Circulation, 109*(16), 2031–2041. doi:10.1161/01.CIR.0000126280.65777.A4 PMID:15117863

Harden, R. N. (2005). Pharmacotherapy of complex regional pain syndrome. *American Journal of Physical Medicine & Rehabilitation, 84*(3), S17–S28. doi:10.1097/01.PHM.0000154906.87708.C9 PMID:15722780

Harden, R. N., Oaklander, A. L., Burton, A. W., Perez, R. S., Richardson, K., Swan, M., Barthel, J., Costa, B., Graciosa, J. R., & Bruehl, S. (2013). Complex regional pain syndrome: Practical diagnostic and treatment guidelines. *Pain Medicine, 14*(2), 180–229. doi:10.1111/pme.12033 PMID:23331950

Harrison, R. A., & Field, T. S. (2015). Post stroke pain: Identification, assessment, and therapy. *Cerebrovascular Diseases (Basel, Switzerland), 39*(3-4), 190–201. doi:10.1159/000375397 PMID:25766121

Hecht, J. S. (1992). Subscapular nerve block in the painful hemiplegic shoulder. *Archives of Physical Medicine and Rehabilitation, 73*(11), 1036–1039. PMID:1444768

Indredavik, B., Rohweder, G., Naalsund, E., & Lydersen, S. (2008). Medical complications in a comprehensive stroke unit and an early supported discharge service. *Stroke, 39*(2), 414–420. doi:10.1161/STROKEAHA.107.489294 PMID:18096834

Iurato, L., Randisi, G., Torrisi, J., Grimaldi, R., Naso, G., Verniccio, S., Occhipinti, C., & Vecchio, M. M. (2015). A temible complication of ischemic stroke: Pulmonary embolism. *Clinical Management Issues, 9*(4), 101–108. doi:10.7175/cmi.v9i4.1196

Johnston, K. C., Li, J. Y., Lyden, P. D., Hanson, S. K., Feasby, T. E., Adams, R. J., Faught, R. E. Jr, & Haley, E. C. Jr. (1998). Medical and neurological complications of ischemic stroke: Experience from the RANTTAS trial. *Stroke, 29*(2), 447–453. doi:10.1161/01.STR.29.2.447 PMID:9472888

Jönsson, A.-C., Lindgren, I., Hallström, B., Norrving, B., & Lindgren, A. (2006). Prevalence and intensity of pain after stroke: A population based study focusing on patients' perspectives. *Journal of Neurology, Neurosurgery, and Psychiatry*, *77*(5), 590–595. doi:10.1136/jnnp.2005.079145 PMID:16354737

Kelly, J., Rudd, A., Lewis, R., & Hunt, B. (2001). Venous thromboembolism after acute stroke. *Stroke*, *32*(1), 262–267. doi:10.1161/01.STR.32.1.262 PMID:11136946

Kendall, R. (2010). Musculoskeletal problems in stroke survivors. *Topics in Stroke Rehabilitation*, *17*(3), 173–178. doi:10.1310/tsr1703-173 PMID:20797960

Khan, M. T., Ikram, A., Saeed, O., Afridi, T., Sila, C. A., Smith, M. S., ... Shuaib, A. (2017). Deep vein thrombosis in acute stroke-a systemic review of the literature. *Cureus*, *9*(12). Advance online publication. doi:10.7759/cureus.1982 PMID:29503776

Koog, Y. H., Jin, S. S., Yoon, K., & Min, B.-I. (2010). Interventions for hemiplegic shoulder pain: Systematic review of randomised controlled trials. *Disability and Rehabilitation*, *32*(4), 282–291. doi:10.3109/09638280903127685 PMID:20055567

Kumar, S., Selim, M. H., & Caplan, L. R. (2010). Medical complications after stroke. *Lancet Neurology*, *9*(1), 105–118. doi:10.1016/S1474-4422(09)70266-2 PMID:20083041

Lewis, R. P., & Kriukelyte, I. (2019). Complex neuropathic pain states. *Anaesthesia and Intensive Care Medicine*, *20*(10), 579–582. doi:10.1016/j.mpaic.2019.07.013

Mari, F., Matei, M., Ceravolo, M. G., Pisani, A., Montesi, A., & Provinciali, L. (1997). Predictive value of clinical indices in detecting aspiration in patients with neurological disorders. *Journal of Neurology, Neurosurgery, and Psychiatry*, *63*(4), 456–460. doi:10.1136/jnnp.63.4.456 PMID:9343123

McKenna, L. B. (2001). Hemiplegic shoulder pain: Defining the problem and its management. *Disability and Rehabilitation*, *23*(16), 698–705. doi:10.1080/09638280110062149 PMID:11732559

Mehdi, Z., Birns, J., & Bhalla, A. (2013). Post-stroke urinary incontinence. *International Journal of Clinical Practice*, *67*(11), 1128–1137. doi:10.1111/ijcp.12183 PMID:23834208

National Institute for Health Care Excellence. (2018). *Venous Thromboembolism in over 16s: reducing the risk of hospital-acquired deep vein thrombosis or pulmonary embolism*. NICE Guideline [NG89].

Nelson, L. A. (2007). The role of biofeedback in stroke rehabilitation: Past and future directions. *Topics in Stroke Rehabilitation*, *14*(4), 59–66. doi:10.1310/tsr1404-59 PMID:17698458

Osawa, A., Maeshima, S., & Tanahashi, N. (2013). Water-swallowing test: Screening for aspiration in stroke patients. *Cerebrovascular Diseases (Basel, Switzerland)*, *35*(3), 276–281. doi:10.1159/000348683 PMID:23548854

Ostaszkiewicz, J., Johnston, L., & Roe, B. (2004). Timed voiding for the management of urinary incontinence in adults. *Cochrane Database of Systematic Reviews*, (1), 002802. doi:10.1002/14651858.CD002802.pub2 PMID:14973993

Paci, M., Nannetti, L., & Rinaldi, L. A. (2005). Glenohumeral subluxation in hemiplegia: An overview. *Journal of Rehabilitation Research and Development*, *42*(4), 557. Advance online publication. doi:10.1682/JRRD.2004.08.0112 PMID:16320150

Paolucci, S., Antonucci, G., Grasso, M. G., Morelli, D., Troisi, E., Coiro, P., De Angelis, D., Rizzi, F., & Bragoni, M. (2001). Post-stroke depression, antidepressant treatment and rehabilitation results. *Cerebrovascular Diseases (Basel, Switzerland)*, *12*(3), 264–271. doi:10.1159/000047714 PMID:11641594

Pretorius, S., & Van Zyl, L. (2008). Nutritional therapy in stroke management: prevention and rehabilitation: nutrition. *CME: Your SA Journal of CPD, 26*(3), 164-166. https://hdl.handle.net/10520/EJC63446

Ramsey, D., Smithard, D., & Kalra, L. (2005). Silent aspiration: What do we know? *Dysphagia*, *20*(3), 218–225. doi:10.100700455-005-0018-9 PMID:16362510

Ramsey, D. J., Smithard, D. G., & Kalra, L. (2003). Early assessments of dysphagia and aspiration risk in acute stroke patients. *Stroke*, *34*(5), 1252–1257. doi:10.1161/01.STR.0000066309.06490.B8 PMID:12677020

Robinson, R. G., & Jorge, R. E. (2016). Post-stroke depression: A review. *The American Journal of Psychiatry*, *173*(3), 221–231. doi:10.1176/appi.ajp.2015.15030363 PMID:26684921

Robinson, R. G., & Spalletta, G. (2010). Poststroke depression: A review. *Canadian Journal of Psychiatry*, *55*(6), 341–349. doi:10.1177/070674371005500602 PMID:20540828

Scheitz, J. F., Nolte, C. H., Doehner, W., Hachinski, V., & Endres, M. (2018). Stroke–heart syndrome: Clinical presentation and underlying mechanisms. *Lancet Neurology*, *17*(12), 1109–1120. doi:10.1016/S1474-4422(18)30336-3 PMID:30509695

Shen, S.-X., & Liu, Y. (2018). A retrospective study of neuromuscular electrical stimulation for treating women with post-stroke incontinence. *Medicine*, *97*(26), e11264. Advance online publication. doi:10.1097/MD.0000000000011264 PMID:29952999

Shi, Y., Yang, D., Zeng, Y., & Wu, W. (2017). Risk factors for post-stroke depression: A meta-analysis. *Frontiers in Aging Neuroscience*, *9*, 218. doi:10.3389/fnagi.2017.00218 PMID:28744213

Smith, H. A., Lee, S. H., O'Neill, P. A., & Connolly, M. J. (2000). The combination of bedside swallowing assessment and oxygen saturation monitoring of swallowing in acute stroke: A safe and humane screening tool. *Age and Ageing*, *29*(6), 495–499. doi:10.1093/ageing/29.6.495 PMID:11191240

Smithard, D., O'neill, P., Park, C., Morris, J., Wyatt, R., England, R., & Martin, D. (1996). Complications and outcome after acute stroke: Does dysphagia matter? *Stroke*, *27*(7), 1200–1204. doi:10.1161/01.STR.27.7.1200 PMID:8685928

Sommerfeld, D. K., Gripenstedt, U., & Welmer, A.-K. (2012). Spasticity after stroke: An overview of prevalence, test instruments, and treatments. *American Journal of Physical Medicine & Rehabilitation*, *91*(9), 814–820. doi:10.1097/PHM.0b013e31825f13a3 PMID:22760104

Sura, L., Madhavan, A., Carnaby, G., & Crary, M. A. (2012). Dysphagia in the elderly: Management and nutritional considerations. *Clinical Interventions in Aging*, *7*, 287. doi:10.2147/CIA.S23404 PMID:22956864

Teasell, R., Foley, N., Pereira, S., Sequeira, K., & Miller, T. (2012). Evidence to practice: Botulinum toxin in the treatment of spasticity post stroke. *Topics in Stroke Rehabilitation*, *19*(2), 115–121. doi:10.1310/tsr1902-115 PMID:22436359

Terre, R., & Mearin, F. (2006). Oropharyngeal dysphagia after the acute phase of stroke: Predictors of aspiration. *Neurogastroenterology and Motility*, *18*(3), 200–205. doi:10.1111/j.1365-2982.2005.00729.x PMID:16487410

Thibaut, A., Chatelle, C., Ziegler, E., Bruno, M.-A., Laureys, S., & Gosseries, O. (2013). Spasticity after stroke: Physiology, assessment and treatment. *Brain Injury: [BI]*, *27*(10), 1093–1105. doi:10.3109/02699052.2013.804202 PMID:23885710

Turner-Stokes, L., & Jackson, D. (2002). Shoulder pain after stroke: A review of the evidence base to inform the development of an integrated care pathway. *Clinical Rehabilitation*, *16*(3), 276–298. doi:10.1191/0269215502cr491oa PMID:12017515

Vafadar, A. K., Côté, J. N., & Archambault, P. S. (2015). Effectiveness of functional electrical stimulation in improving clinical outcomes in the upper arm following stroke: A systematic review and meta-analysis. *BioMed Research International*, *2015*, 1–14. Advance online publication. doi:10.1155/2015/729768 PMID:25685805

Van Ouwenaller, C., Laplace, P., & Chantraine, A. (1986). Painful shoulder in hemiplegia. *Archives of Physical Medicine and Rehabilitation*, *67*(1), 23–26. PMID:3942479

Villa, R. F., Ferrari, F., & Moretti, A. (2018). Post-stroke depression: Mechanisms and pharmacological treatment. *Pharmacology & Therapeutics*, *184*, 131–144. doi:10.1016/j.pharmthera.2017.11.005 PMID:29128343

Walsh, K. (2001). Management of shoulder pain in patients with stroke. *Postgraduate Medical Journal*, *77*(912), 645–649. doi:10.1136/pmj.77.912.645 PMID:11571371

Yu, D. (2004). Shoulder pain in hemiplegia. *Physical Medicine and Rehabilitation Clinics of North America*, *15*(3), 683–697. doi:10.1016/S1047-9651(03)00130-X PMID:15219895

Zorowitz, R. D., & Tietjen, G. E. (1999). Medical complications after stroke. *Journal of Stroke and Cerebrovascular Diseases*, *8*(3), 192–196. doi:10.1016/S1052-3057(99)80026-3 PMID:17895163

# Chapter 10
# Transient Ischemic Attack

**Murat Guntel**
*Mustafa Kemal University, Turkey*

## ABSTRACT

*The transient ischemic attack is a neurological emergency which is a clinical view of focal cerebral, retinal, or spinal dysfunction that lasts less than an hour, without any detectable acute infarction in neurological imaging methods. TIA is a serious warning for ischemic stroke, and this risk is particularly high in the first 48 hours. Following TIAs, approximately 10-15% of patients undergo stroke in 90 days and about half of these patients suffer a stroke in the first two days. Neuroimaging and laboratory studies should be performed quickly to reveal the etiology and to reduce the risk of stroke that may develop in patients present with TIA. Therapeutic and preventive interventions should be started as soon as possible. With early diagnosis and treatment, the risk of a 90-day stroke in these patients can be reduced by 80%. In addition to antiplatelet and anticoagulant treatments, aggressive control of blood pressure, regulation of blood sugar, statin, dietary recommendations, exercise, and managing the other underlying specific conditions should be started quickly.*

## INTRODUCTION

According to the World Health Organization (WHO), Transient Ischemic Attack (TIA), is defined as focal neurological symptoms that are thought to be of vascular origin lasting less than 24 hours (Sorensen & Ay, 2011). This definition was updated by American Heart and Stroke Society (Association) (AHA / ASA) in 2009 as focal cerebral, spinal and retinal ischemia, which lasted less than one hour, in which infarction could not be detected by neurological imaging methods (Easton et al., 2009). TIA typically lasts less than an hour and often lasts within minutes (in only

DOI: 10.4018/978-1-7998-7433-1.ch010

14% of cases, the clinical picture lasted longer than 6 hours). A few second episodes are unlikely to be due to TIA, but patients with lesions detected in diffusion MRI are defined as ischemic stroke, even if their symptoms improve in a very short time (Easton et al., 2009).

TIAs are often recurrent and should not be permanent neurological deficits. TIA findings occur due to sudden changes in vascular dynamics. It is more appropriate to consider this clinic view as a stroke if the symptoms last more than an hour. If the neurological deficit persists for more than 1 hour, the chance of recovery spontaneously is less than 15%. Modern imaging methods have shown the presence of ischemic lesions in the brain for the majority of transient symptoms that last more than an hour. This condition is called ''Transient Symptoms with Infarction''. In addition to transient neurological deficits, the presence of ischemic lesions in imaging is typical in this picture (Ay et al., 2005; Levy, 1988; Nazliel, 2018 ; Siket & Edlow, 2013).

## CLINICAL FINDINGS AND RISK FACTORS

TIAs are divided into two main groups as "carotid system" and "vertebrobasilar system" according to the vascular areas reflected by the symptoms. Typical symptoms for TIA include: unilateral face, arm, leg weakness, unilateral sensory loss, dysphasia-aphasia, imbalance, vertigo, diplopia, amaurosis fugax, and rarely hemianopia (Nadarajan, Perry, Johnson, & Werring, 2014). Symptoms such as generalized weakness or generalized sensory loss, confusion, consciousness disorders, syncope, dizziness, incontinence, amnesia, positive symptoms such as involuntary movement, tingling, seeing flashing lights, and bilateral blurred vision are not considered as typical symptoms for TIA (Coutts, 2017).

Pathophysiological mechanisms of TIA are like ischemic stroke. Main mechanisms are transient interruption of a certain artery in the related brain area, differences in cerebral hypoperfusion tolerability, collateral flow, and oxygen-carrying capacity of the patient. This carrying capacity is considered to be low in patients with periventricular and subcortical white matter disease (Leukariosis) (Arsava et al., 2011; Dirnagl, Becker, & Meisel, 2009; Nazliel, 2018 ; Siket & Edlow, 2013).

The main risk factors for the development of TIA include: Age, hypertension (HT), hypercholesterolemia, diabetes mellitus (DM), atrial fibrillation (AF), previous myocardial infarction (MI), carotid artery stenosis, cardiomyopathy, nutritional habits, physical inactivation, smoking, excessive alcohol use and psychosocial stress (Easton et al., 2009; Khare, 2016). Following TIAs, approximately 10-15% of patients undergo stroke in 90 days and about half of these patients suffer a stroke in the first two days (Easton et al., 2009). The risk of having a stroke in the first 24 hours after TIA is approximately 4%. This rate is 2 times the risk of developing

MI or death in patients presenting with acute coronary syndrome (2% in 24 hours) (Johnston, Gress, Browner, & Sidney, 2000; Lovett et al., 2003; Nazliel, 2018).

## DIAGNOSTIC METHODS

History of current complaint, physical and neurological examination, biochemical-hematological evaluation should be completed quickly in the diagnosis stage of patients with TIA. The onset of complaints, their duration, aggravating and relieving conditions, accompanying diseases, medications used, previous TIA, MI, or stroke history should be questioned in the anamnesis. Cardiac auscultation, carotid artery auscultation, and blood pressure measurement should be included in the physical examination. A complete neurological examination should be performed, especially for common symptoms, focal neurological deficits, and speech disturbances. Computed tomography (CT), magnetic resonance imaging (MRI), also, vascular examination methods such as Computed Tomography Angiography (CTA), Magnetic Resonance Angiography (MRA), Doppler ultrasonography (USG) should be performed quickly, whichever is available at the center. A detailed examination should be made in terms of cardiac sources of embolism. All patients should be evaluated in detail in terms of murmur in the neck vessels and heart, decreased or lost peripheral pulses, increased/ decreased blood pressure, rhythm disturbance / atrial fibrillation (Edlow, 2018).

AHA / ASA recommends that patients be imaged, preferably by diffusion-weighted imaging, within the first 24 hours after symptoms (Easton et al., 2009). Approximately 50-70% of the patients who apply to the emergency department with TIA are examined with CT (Edlow, Kim, Pelletier, & Camargo, 2006; Johnston et al., 2003). In TIA patients, brain and diffusion MRI is highly sensitive compared to CT, especially in detecting small infarcts. CT and, if possible, CT angiography is recommended for patients who cannot undergo MRI.

Cervicocephalic vasculature must be evaluated in terms of the presence of atherosclerotic lesions and the most preferred methods in noninvasive vascular imaging are BTA, MRA, and Doppler USG. In patients planned for carotid endarterectomy, carotid imaging should be performed within 1 week after the onset of symptoms.

Doppler USG is an inexpensive technique, with no contrast agent and no radiation however, it depends on the practitioner's experience. If it is not performed with transcranial doppler ultrasound (TCD), it does not give information about intracranial vessels (Easton et al., 2009; Heijenbrok-Kal, Buskens, Nederkoorn, van der Graaf, & Hunink, 2006).

BTA is a rapid examination method, it can be performed 24/7 in many centers. All craniocervical vessels can be imaged with the use of one dose contrast agent. Although dissection sensitivity is high, radiation risk and contrast nephrotoxicity limits its use (Josephson et al., 2004). MRA has 70-99% sensitivity in carotid stenosis. It is the examination that is considered to have the highest sensitivity compared to USG and CT (Wardlaw et al., 2006). Time of Flight (TOF) images is an important alternative for patients who cannot be used gadolinium, but it is an expensive, time-consuming examination and its usage is limited (Nazliel, 2018; Phan, Huston, Bernstein, Riederer, & Brown, 2001).

Cardiac evaluation must be performed to detect the presence of a source of cardioembolism such as patent foramen ovale, valvular disease, cardiac thrombus, and atherosclerosis. Also, the risk of cardiac events increased in patients with TIA. 3% of patients with TIA were hospitalized within 90 days due to myocardial infarction, unstable angina, or ventricular arrhythmias, so it is recommended that patients be examined quickly by electrocardiography (ECG), echocardiography (ECHO)/ transesophageal echo (TEE). Some centers recommend patients with ischemic stroke and TIA to be evaluated for at least 24 hours with in-hospital ECG or 7 days with continuous monitoring outside the hospital. It is known that monitoring for more than 24 hours significantly increases the rate of AF diagnosis (Verma et al., 2014).

## RISK STRATIFICATION TOOLS FOR TRANSIENT ISCHEMIC ATTACK

Several scoring systems have been developed to determine the risk of stroke in patients with TIA. The most common of these is the ABCD2 score (Table 1) (Rothwell et al., 2005). In this scoring system, patients are scored based on the presence of the following features: A (Age), B (Blood Pressure), C (Features) D2 (Duration-Diabetes). The total score is 7. While those scoring between 0-3 constitute the low-risk group, the risk of stroke is considered to be around 1% in these patients over 2 days. This risk is considered to be 4.1% in the medium-risk group with 4-5 points and approximately 8.1% in the high-risk group with 6-7 points (Amarenco et al., 2016; Bhatt & Jani, 2011; Johnston et al., 2007). Most stroke centers will hospitalize patients with TIA for accelerated evaluation and management if the ABCD2 score is 4 or higher. ABCD2 score is criticized for some aspects. It is thought to identify patients with high disability due to stroke rather than stroke risk. Also, it does not include imaging features. It contains no evidence of etiology and it is considered to be inadequate in determining processes related to large vessel atherosclerosis (Chang et al., 2018; Nazliel, 2018).

Merwick et al in 2010 described 2 new scoring systems to predict stroke risk, ABCD3 and ABCD3-I (Merwick et al., 2010). In the ABCD3 scoring system, dual TIA (the presence of ≥2 TIA symptoms within 7 days) is added to the ABCD2 score. Abnormal findings detected in neuroimaging (such as carotid stenosis or acute diffusion-weighted imaging hyperintensity) were also added to the ABCD3-I scoring system (Kiyohara et al., 2014; Merwick et al., 2010). In this new scoring system, having multiple attacks, the presence of large vessel atherosclerosis in etiology, and detection of diffusion restricting lesions in MRI is added next to the features in the ABCD2 score. The presence of each pathology is scored with 2 points, and in this system, the total score is 13. In this scoring system, 0-3 score is low, 4-7 score is medium and 8-13 score is a high-risk group. This scoring system is considered to be more sensitive than the ABCD2 score in predicting stroke risk. In the differential diagnosis of TIA, seizure, hypotension, hypo/hyperglycemia, drop attacks, peripheral vestibular disorder, migraine with aura, chronic subdural hematoma, structural intracranial lesions, and space-occupying lesions should be considered (Chang et al., 2018; Nazliel, 2018).

*Table 1. ABCD2 score*

| Risk Factor | Points |
|---|---|
| Age ≥60 years | 1 |
| Blood Pressure ≥ 140/90 mmHg | 1 |
| Clinical Features: Unilateral weakness<br>Clinical Features: Speech disturbance without weakness | 2<br>1 |
| Duration of symptoms: ≥ 60 minutes<br>Duration of symptoms:10-59 minutes | 2<br>1 |
| Diabetes Mellitus | 1 |
| **Total** | **0-7** |

*Table 2. ABCD3-I score*

| | |
|---|---|
| Age ≥60 y | 1 |
| Blood pressure ≥140/90 mmHg | 1 |
| Clinical features: Unilateral weakness<br>Clinical features: Speech impairment without weakness | 2<br>1 |
| Duration of symptoms: ≥ 60 minutes<br>Duration of symptoms:10-59 minutes | 2<br>1 |
| Diabetes Mellitus | 1 |
| Dual TIA | 2 |
| Imaging: Ipsilateral ≥50% stenosis of internal carotid artery | 2 |
| Imaging: Acute diffusion-weighted imaging hyperintensity | 2 |
| **Total** | **0-13** |

# TREATMENT AND MANAGEMENT

## General Principles

The main goal in the TIA treatment approach is to reduce the risk of subsequent stroke and the risk of stroke is very high in the first week after TIA, but it can be reduced by 80% with early-effective medical treatment and surgical and endovascular approaches when necessary. Specific treatment approaches for etiological reasons are required in patients. It is difficult to complete all etiological evaluation in emergency departments. Apart from routine examinations, oxygenation, blood pressure control, and blood glucose regulation should be provided quickly in the emergency departments (Kennedy et al., 2007).

## Antiplatelet Therapy

Early antithrombotic treatment in patients with TIA leads to an 80% relativistic risk reduction. CHANCE trial and the multinational POINTE trials showed that the use of dual antiplatelet (aspirin and clopidogrel) for 3 weeks to 1 month and then single antiplatelet agent use is the most appropriate antiaggregant treatment scheme. The Fast Assessment of Stroke and Transient Ischemic Attack to prevent Early Recurrence (FASTER) study revealed that the combination of aspirin and clopidogrel in patients with TIA in the hyperacute period is an effective approach (Kennedy et al., 2007; Wang et al., 2015). It is known that patients who are followed up with the diagnosis of large vessel atherosclerosis benefit from dual antiaggregant therapy (Hankey et al., 2011). The Triple Antiplatelets for Reducing Dependency after Ischemic Stroke (TARDIS) study revealed that triple therapy applied by adding aspirin and clopidogrel to dipyridamole was not effective on the risk of stroke (Krishnan et al., 2015).

Other antiplatelet agents have been emphasized in secondary prevention of stroke in the past few years. Although cilostazol, a similar agent to dipyridamole, which inhibits phosphodiesterase and increases cyclic AMP concentrations, has been shown superior to aspirin in reducing vascular events in two studies conducted in Asia, Also, cilostazol seemed to cause less bleeding complications, and had more mild adverse effects than aspirin but, it is not known whether it has the same effect on other populations (Kamal, Naqvi, Husain, & Khealani, 2011; Kim, Kwon, & Uchiyama, 2015). The SOCRATES study revealed that the PY2 inhibitor ticagrelor is more effective than aspirin and at least as reliable as aspirin in reducing major events in patients with TIA and atherosclerosis (Amarenco et al., 2017).

## Anticoagulant Therapy

Oral anticoagulation is recommended for patients with atrial fibrillation or other sources of cardioembolic TIA. Warfarin, a vitamin K antagonist, is widely used in atrial fibrillation, mechanical heart valve prosthesis, and cardiogenic embolic events. Studies have shown that anticoagulants are more effective than aspirin in preventing embolisms in patients with AF (Hart, Pearce, & Aguilar, 2007; Zhang, Chen, & Zhang, 2015).

In recent years, direct thrombin inhibitor dabigatran and factor Xa inhibitors rivaroxaban, edoxaban, apixaban have been widely used in non-valvular AF patients. Short half-lives, fewer drug interactions, and no need for laboratory monitoring are the main advantages against warfarin. These drugs are excreted by the kidneys and should be monitored closely in patients who are followed up with the diagnosis of renal failure (Nazliel, 2018).

## Arterial Revascularization

It is known that patients with more than 50% stenosis due to carotid atherosclerosis benefit from carotid endarterectomy or carotid artery stenting. Both techniques, if possible, should be applied in experienced centers within the first two weeks and surgeon's complication rates are also very important regarding which method should be preferred (Edlow, 2018; Kernan et al., 2014).

## Modification of Vascular Risk Factors

Studies have shown that appropriate diet, regular exercises, antiplatelet, statin and antihypertensive therapy combinations reduce the risk of stroke by 80-90% (Hackam & Spence, 2007).

It is important to keep blood pressure under control in TIA patients to reduce the risk of stroke and the risk of TIA recurrence. The question of what the ideal blood pressure should be in the first few hours after the event is uncertain, and it is thought that reducing it to normal values may aggravate ischemia. Rapid initiation of antihypertensive therapy in the acute phase of ischemic stroke is indicated only for patients who are planned to be treated with recombinant tissue-type plasminogen activator or patients with extremely high blood pressure values. Blood pressure should be <185/110 mm / Hg in patients before initiating IV thrombolysis treatment and should be kept at <180/105 mm / Hg for 24 hours after treatment. The benefits of acute lowering of blood pressure in patients with TIA or stroke who will not receive thrombolytic therapy are unclear. In patients with systolic blood pressure> 220 mm/Hg, diastolic blood pressure> 120 mm/Hg, or any other definitive indication, antihypertensive therapy is recommended in the acute phase. Rapid lowering of blood pressure can also be harmful. Daily reductions of 15-20% are recommended (Kernan et al., 2014; Wajngarten & Silva, 2019).

Intensive cholesterol-lowering therapy is recommended for patients who have recently had a stroke or TIA. Studies have shown that patients who had a target LDL cholesterol level of less than 70 mg/dl had a lower risk of subsequent cardiovascular events than those who had a target range of 90-110 g/dl (Amarenco et al., 2020).

Diabetes and metabolic syndrome increase the risk of stroke. Hb A1c levels and body mass indexes in patients with TIA should be monitored (Kernan et al., 2014). It is known that impaired glycemic control negatively affects the effectiveness of clopidogrel. Right diet and exercise plans should be given to these patients, and fasting blood glucose levels should be reduced to <126 mg / dL with oral antidiabetic agents and insülin (Kernan et al., 2014; Lin et al., 2017). There are different pharmacological agents used to control blood glucose. The Insulin Resistance Intervention after Stroke (IRIS) study revealed that pioglitazone reduced the risk of stroke and myocardial infarction by 24% in patients with TIA (Kernan et al., 2016; Nazliel, 2018).

Also, necessary suggestions and warnings should be made about risk-reducing factors such as adequate nutrition, smoking cessation, regular exercise, appropriate body mass index, and limitation of alcohol consumption.

## CONCLUSİON

20% of strokes develop following TIA. Most of the TIAs are not recognized by the patient and therefore patients do not apply to the doctor. 85% of strokes that develop following TIA create disability or have a high mortality rate. Early and effective medical treatment and surgical interventions such as endarterectomy can reduce the 90-day stroke risk that develops after TIA by 80%. It should be kept in mind that these patients are also at high risk for myocardial infarction and cardiovascular events besides stroke (Arsava et al., 2011; Rothwell et al., 2007).

## REFERENCES

Amarenco, P., Albers, G. W., Denison, H., Easton, J. D., Evans, S. R., Held, P., Hill, M. D., Jonasson, J., Kasner, S. E., Ladenvall, P., Minematsu, K., Molina, C. A., Wang, Y., Wong, K. S. L., & Johnston, S. C. (2017). Efficacy and safety of ticagrelor versus aspirin in acute stroke or transient ischaemic attack of atherosclerotic origin: A subgroup analysis of SOCRATES, a randomised, double-blind, controlled trial. *Lancet Neurology*, *16*(4), 301–310. doi:10.1016/S1474-4422(17)30038-8 PMID:28238711

Amarenco, P., Kim, J. S., Labreuche, J., Charles, H., Abtan, J., Béjot, Y., Cabrejo, L., Cha, J.-K., Ducrocq, G., Giroud, M., Guidoux, C., Hobeanu, C., Kim, Y.-J., Lapergue, B., Lavallée, P. C., Lee, B.-C., Lee, K.-B., Leys, D., Mahagne, M.-H., ... Bruckert, E. (2020). A Comparison of Two LDL Cholesterol Targets after Ischemic Stroke. *The New England Journal of Medicine*, *382*(1), 18. doi:10.1056/NEJMoa1910355 PMID:31738483

Amarenco, P., Lavallée, P. C., Labreuche, J., Albers, G. W., Bornstein, N. M., Canhão, P., Caplan, L. R., Donnan, G. A., Ferro, J. M., Hennerici, M. G., Molina, C., Rothwell, P. M., Sissani, L., Školoudík, D., Steg, P. G., Touboul, P.-J., Uchiyama, S., Vicaut, É., & Wong, L. K. (2016). One-Year Risk of Stroke after Transient Ischemic Attack or Minor Stroke. *The New England Journal of Medicine*, *374*(16), 1533–1542. doi:10.1056/NEJMoa1412981 PMID:27096581

Arsava, E. M., Bayrlee, A., Vangel, M., Rost, N. S., Rosand, J., Furie, K. L., Sorensen, A. G., & Ay, H. (2011). Severity of leukoaraiosis determines clinical phenotype after brain infarction. *Neurology*, *77*(1), 55–61. doi:10.1212/WNL.0b013e318221ad02 PMID:21700580

Ay, H., Koroshetz, W. J., Benner, T., Vangel, M. G., Wu, O., Schwamm, L. H., & Sorensen, A. G. (2005). Transient ischemic attack with infarction: A unique syndrome? *Annals of Neurology*, *57*(5), 679–686. doi:10.1002/ana.20465 PMID:15852402

Bhatt, A., & Jani, V. (2011). The ABCD and ABCD2 Scores and the Risk of Stroke following a TIA: A Narrative Review. *ISRN Neurology*, *518621*(10), 21. doi:10.5402/2011/518621 PMID:22389822

Chang, B. P., Rostanski, S., Willey, J., Kummer, B., Miller, E., & Elkind, M. (2018). Can I Send This Patient with Stroke Home? Strategies Managing Transient Ischemic Attack and Minor Stroke in the Emergency Department. *The Journal of Emergency Medicine*, *54*(5), 636–644. doi:10.1016/j.jemermed.2017.12.015 PMID:29321107

Coutts, S. B. (2017). Diagnosis and Management of Transient Ischemic Attack. *Continuum, 23*(1), 82-92.

Dirnagl, U., Becker, K., & Meisel, A. (2009). Preconditioning and tolerance against cerebral ischaemia: From experimental strategies to clinical use. *Lancet Neurology*, *8*(4), 398–412. doi:10.1016/S1474-4422(09)70054-7 PMID:19296922

Easton, J. D., Saver, J. L., Albers, G. W., Alberts, M. J., Chaturvedi, S., Feldmann, E., Hatsukami, T. S., Higashida, R. T., Johnston, S. C., Kidwell, C. S., Lutsep, H. L., Miller, E., & Sacco, R. L. (2009). Definition and evaluation of transient ischemic attack: A scientific statement for healthcare professionals from the American Heart Association/American Stroke Association Stroke Council; Council on Cardiovascular Surgery and Anesthesia; Council on Cardiovascular Radiology and Intervention; Council on Cardiovascular Nursing; and the Interdisciplinary Council on Peripheral Vascular Disease. The American Academy of Neurology affirms the value of this statement as an educational tool for neurologists. *Stroke*, *40*(6), 2276–2293. doi:10.1161/STROKEAHA.108.192218 PMID:19423857

Edlow, J. A. (2018). Managing Patients With Transient Ischemic Attack. *Annals of Emergency Medicine*, *71*(3), 409–415. doi:10.1016/j.annemergmed.2017.06.026 PMID:28754355

Edlow, J. A., Kim, S., Pelletier, A. J., & Camargo, C. A. Jr. (2006). National study on emergency department visits for transient ischemic attack, 1992-2001. *Academic Emergency Medicine*, *13*(6), 666–672. PMID:16609106

Hackam, D. G., & Spence, J. D. (2007). Combining multiple approaches for the secondary prevention of vascular events after stroke: A quantitative modeling study. *Stroke*, *38*(6), 1881–1885. doi:10.1161/STROKEAHA.106.475525 PMID:17431209

Hankey, G. J., Johnston, S. C., Easton, J. D., Hacke, W., Mas, J. L., Brennan, D., Mak, K. H., Bhatt, D. L., Fox, K. A. A., & Topol, E. J. (2011). Effect of clopidogrel plus ASA vs. ASA early after TIA and ischaemic stroke: A substudy of the CHARISMA trial. *International Journal of Stroke*, *6*(1), 3–9. doi:10.1111/j.1747-4949.2010.00535.x PMID:21205234

Hart, R. G., Pearce, L. A., & Aguilar, M. I. (2007). Meta-analysis: Antithrombotic therapy to prevent stroke in patients who have nonvalvular atrial fibrillation. *Annals of Internal Medicine*, *146*(12), 857–867. doi:10.7326/0003-4819-146-12-200706190-00007 PMID:17577005

Heijenbrok-Kal, M. H., Buskens, E., Nederkoorn, P. J., van der Graaf, Y., & Hunink, M. G. (2006). Optimal peak systolic velocity threshold at duplex us for determining the need for carotid endarterectomy: A decision analytic approach. *Radiology*, *238*(2), 480–488. doi:10.1148/radiol.2381041078 PMID:16371575

Johnston, S. C., Fayad, P. B., Gorelick, P. B., Hanley, D. F., Shwayder, P., van Husen, D., & Weiskopf, T. (2003). Prevalence and knowledge of transient ischemic attack among US adults. *Neurology*, *60*(9), 1429–1434. doi:10.1212/01.WNL.0000063309.41867.0F PMID:12743226

Johnston, S. C., Gress, D. R., Browner, W. S., & Sidney, S. (2000). Short-term prognosis after emergency department diagnosis of TIA. *Journal of the American Medical Association, 284*(22), 2901–2906. doi:10.1001/jama.284.22.2901 PMID:11147987

Johnston, S. C., Rothwell, P. M., Nguyen-Huynh, M. N., Giles, M. F., Elkins, J. S., Bernstein, A. L., & Sidney, S. (2007). Validation and refinement of scores to predict very early stroke risk after transient ischaemic attack. *Lancet, 369*(9558), 283–292. doi:10.1016/S0140-6736(07)60150-0 PMID:17258668

Josephson, S. A., Bryant, S. O., Mak, H. K., Johnston, S. C., Dillon, W. P., & Smith, W. S. (2004). Evaluation of carotid stenosis using CT angiography in the initial evaluation of stroke and TIA. *Neurology, 63*(3), 457–460. doi:10.1212/01.WNL.0000135154.53953.2C PMID:15304575

Kamal, A. K., Naqvi, I., Husain, M. R., & Khealani, B. A. (2011). Cilostazol versus aspirin for secondary prevention of vascular events after stroke of arterial origin. *Cochrane Database of Systematic Reviews, 19*(1). PMID:21249700

Kennedy, J., Hill, M. D., Ryckborst, K. J., Eliasziw, M., Demchuk, A. M., & Buchan, A. M. (2007). Fast assessment of stroke and transient ischaemic attack to prevent early recurrence (FASTER): A randomised controlled pilot trial. *Lancet Neurology, 6*(11), 961–969. doi:10.1016/S1474-4422(07)70250-8 PMID:17931979

Kernan, W. N., Ovbiagele, B., Black, H. R., Bravata, D. M., Chimowitz, M. I., Ezekowitz, M. D., Fang, M. C., Fisher, M., Furie, K. L., Heck, D. V., Johnston, S. C. C., Kasner, S. E., Kittner, S. J., Mitchell, P. H., Rich, M. W., Richardson, D. J., Schwamm, L. H., & Wilson, J. A. (2014). Guidelines for the prevention of stroke in patients with stroke and transient ischemic attack: A guideline for healthcare professionals from the American Heart Association/American Stroke Association. *Stroke, 45*(7), 2160–2236. doi:10.1161/STR.0000000000000024 PMID:24788967

Kernan, W. N., Viscoli, C. M., Furie, K. L., Young, L. H., Inzucchi, S. E., Gorman, M., Guarino, P. D., Lovejoy, A. M., Peduzzi, P. N., Conwit, R., Brass, L. M., Schwartz, G. G., Adams, H. P. Jr, Berger, L., Carolei, A., Clark, W., Coull, B., Ford, G. A., Kleindorfer, D., ... Winder, T. R. (2016). Pioglitazone after Ischemic Stroke or Transient Ischemic Attack. *The New England Journal of Medicine, 374*(14), 1321–1331. doi:10.1056/NEJMoa1506930 PMID:26886418

Khare, S. (2016). Risk factors of transient ischemic attack: An overview. *Journal of Midlife Health, 7*(1), 2–7. doi:10.4103/0976-7800.179166 PMID:27134474

Kim, J. S., Kwon, S. U., & Uchiyama, S. (2015). Cilostazol research in Asia: Can it be applied to European and American patients? *International Journal of Stroke, 1*(1_suppl), 1–9. doi:10.1111/ijs.12460 PMID:25736819

Kiyohara, T., Kamouchi, M., Kumai, Y., Ninomiya, T., Hata, J., Yoshimura, S., ... Kitazono, T. (2014). ABCD3 and ABCD3-I scores are superior to ABCD2 score in the prediction of short- and long-term risks of stroke after transient ischemic attack. *Stroke*, *45*(2), 418–425. doi:10.1161/STROKEAHA.113.003077 PMID:24335223

Krishnan, K., Beridze, M., Christensen, H., Dineen, R., Duley, L., Heptinstall, S., ... Bath, P. (2015). Safety and efficacy of intensive vs. guideline antiplatelet therapy in high-risk patients with recent ischemic stroke or transient ischemic attack: Rationale and design of the Triple Antiplatelets for Reducing Dependency after Ischaemic Stroke (TARDIS) trial (ISRCTN47823388). *International Journal of Stroke*, *10*(7), 1159–1165. doi:10.1111/ijs.12538 PMID:26079743

Levy, D. E. (1988). How transient are transient ischemic attacks? *Neurology*, *38*(5), 674–677. doi:10.1212/WNL.38.5.674 PMID:3362360

Lin, Y., Wang, A., Li, J., Lin, J., Wang, D., Meng, X., Ou, L., Chen, W., Zhao, X., Liu, L., Wang, Y., & Wang, Y. (2017). Impact of Glycemic Control on Efficacy of Clopidogrel in Transient Ischemic Attack or Minor Stroke Patients With CYP2C19 Genetic Variants. *Stroke*, *48*(4), 998–1004. doi:10.1161/STROKEAHA.116.016463 PMID:28289237

Lovett, J. K., Dennis, M. S., Sandercock, P. A., Bamford, J., Warlow, C. P., & Rothwell, P. M. (2003). Very early risk of stroke after a first transient ischemic attack. *Stroke*, *34*(8), 10. doi:10.1161/01.STR.0000080935.01264.91 PMID:12855835

Merwick, A., Albers, G. W., Amarenco, P., Arsava, E. M., Ay, H., Calvet, D., Coutts, S. B., Cucchiara, B. L., Demchuk, A. M., Furie, K. L., Giles, M. F., Labreuche, J., Lavallée, P. C., Mas, J.-L., Olivot, J. M., Purroy, F., Rothwell, P. M., Saver, J. L., Sheehan, Ó. C., ... Kelly, P. J. (2010). Addition of brain and carotid imaging to the $ABCD^2$ score to identify patients at early risk of stroke after transient ischaemic attack: A multicentre observational study. *Lancet Neurology*, *9*(11), 1060–1069. doi:10.1016/S1474-4422(10)70240-4 PMID:20934388

Nadarajan, V., Perry, R. J., Johnson, J., & Werring, D. J. (2014). Transient ischaemic attacks: Mimics and chameleons. *Practical Neurology*, *14*(1), 23–31. doi:10.1136/practneurol-2013-000782 PMID:24453269

Nazliel, B. (2018). *Transient Ischemic Attack. Gokce M, editör. Stroke – I.* Turkish Clinics.

Phan, T., Huston, J. III, Bernstein, M. A., Riederer, S. J., & Brown, R. D. Jr. (2001). Contrast-enhanced magnetic resonance angiography of the cervical vessels: Experience with 422 patients. *Stroke*, *32*(10), 2282–2286. doi:10.1161/hs1001.096046 PMID:11588314

Rothwell, P. M., Giles, M. F., Chandratheva, A., Marquardt, L., Geraghty, O., Redgrave, J. N., Lovelock, C. E., Binney, L. E., Bull, L. M., Cuthbertson, F. C., Welch, S. J. V., Bosch, S., Carasco-Alexander, F., Silver, L. E., Gutnikov, S. A., & Mehta, Z. (2007). Effect of urgent treatment of transient ischaemic attack and minor stroke on early recurrent stroke (EXPRESS study): A prospective population-based sequential comparison. *Lancet*, *370*(9596), 1432–1442. doi:10.1016/S0140-6736(07)61448-2 PMID:17928046

Rothwell, P. M., Giles, M. F., Flossmann, E., Lovelock, C. E., Redgrave, J. N., Warlow, C. P., & Mehta, Z. (2005). A simple score (ABCD) to identify individuals at high early risk of stroke after transient ischaemic attack. *Lancet*, *366*(9479), 29–36. doi:10.1016/S0140-6736(05)66702-5 PMID:15993230

Siket, M. S., & Edlow, J. (2013). Transient ischemic attack: An evidence-based update. *Emergency Medicine Practice*, *15*(1), 1–26. PMID:23257070

Sorensen, A. G., & Ay, H. (2011). Transient ischemic attack: Definition, diagnosis, and risk stratification. *Neuroimaging Clinics of North America*, *21*(2), 303–313. doi:10.1016/j.nic.2011.01.013 PMID:21640301

Verma, A., Cairns, J. A., Mitchell, L. B., Macle, L., Stiell, I. G., Gladstone, D., McMurtry, M. S., Connolly, S., Cox, J. L., Dorian, P., Ivers, N., Leblanc, K., Nattel, S., & Healey, J. S. (2014). 2014 focused update of the Canadian Cardiovascular Society Guidelines for the management of atrial fibrillation. *The Canadian Journal of Cardiology*, *30*(10), 1114–1130. doi:10.1016/j.cjca.2014.08.001 PMID:25262857

Wajngarten, M., & Silva, G. S. (2019). Hypertension and Stroke: Update on Treatment. *European Cardiology*, *14*(2), 111–115. doi:10.15420/ecr.2019.11.1 PMID:31360232

Wang, Y., Pan, Y., Zhao, X., Li, H., Wang, D., Johnston, S. C., Liu, L., Meng, X., Wang, A., Wang, C., & Wang, C. (2015). Clopidogrel With Aspirin in Acute Minor Stroke or Transient Ischemic Attack (CHANCE) Trial: One-Year Outcomes. *Circulation*, *132*(1), 40–46. doi:10.1161/CIRCULATIONAHA.114.014791 PMID:25957224

Wardlaw, J. M., Chappell, F. M., Stevenson, M., De Nigris, E., Thomas, S., Gillard, J., Berry, E., Young, G., Rothwell, P., Roditi, G., Gough, M., Brennan, A., Bamford, J., & Best, J. (2006). Accurate, practical and cost-effective assessment of carotid stenosis in the UK. *Health Technology Assessment*, *10*(30), 1–182. doi:10.3310/hta10300 PMID:16904049

Zhang, J. T., Chen, K. P., & Zhang, S. (2015). Efficacy and safety of oral anticoagulants versus aspirin for patients with atrial fibrillation: A meta-analysis. *Medicine*, *94*(4), 0000000000000409. doi:10.1097/MD.0000000000000409 PMID:25634169

# Chapter 11
# Trunk in Stroke

**Esra Dogru Huzmeli**
*Health Science Faculty, Physiotherapy and Rehabilitation Department, Hatay Mustafa Kemal University, Hatay, Turkey*

**Ozden Gokcek**
*Health Science Faculty, Physiotherapy and Rehabilitation Department, Hatay Mustafa Kemal University, Hatay, Turkey*

## ABSTRACT

*The trunk is the part of the human body that provides basic mechanical stabilization. It provides strength transmission between the upper and lower body regions. Body control is the ability of the body muscles to maintain the upright posture, to adapt to weight transfers, and to maintain selective trunk and limb movements by maintaining the support surface in static and dynamic postural adjustments. Good proximal trunk control provides better distal limb movements, balance, and functional motion. There are many evaluation methods, devices, and scales for trunk function and performance. 3D kinematic, electromyography, hand-held dynamometer, isokinetic dynamometer, trunk accelerometer are some devices that measure trunk function. The motor assessment scale-trunk subscale, the stroke impairment assessment set-trunk control subscale, trunk control test, trunk impairment scale are the most used scales. This chapter explores the effect of strokes on the trunk.*

## INTRODUCTION

The trunk is the part of the human body that provides basic mechanical stabilization, it provides strength transmission between the upper and lower body regions. Body control; is the ability of the body muscles to maintain the upright posture, to adapt to weight transfers, and to maintain selective trunk and limb movements by maintaining

DOI: 10.4018/978-1-7998-7433-1.ch011

the support surface in static and dynamic postural adjustments. Good proximal trunk control provides better distal limb movements, balance, and functional motion (Karthikbabu et al., 2012).

In hemiplegic patients, the trunk muscles are affected negatively in several ways. With isokinetic dynamometer tests, trunk flexor and extensor muscle strength and peak torque were found to be lower in hemiplegic patients than in healthy individuals, and trunk rotator muscle performance was lower than in healthy individuals, but there was no difference between right and left rotation performance (Tanaka et al., 1998; Tanaka et al., 1997; VanCriekinge et al., 2017). In clinical studies, it was reported that the hemiplegic side trunk lateral flexor muscle strength was lower in hemiplegic cases (VanCriekinge et al., 2017; Bohannon et al., 1995). In electromyography (EMG) studies, it was reported that the activity of the latissimus dorsi, abdominal, and paraspinal muscles decreased in hemiplegic patients during exercise. Hemiplegic patients with all these effects; encounters problems such as decreased ambulatory capacity, stability, weight transfer, and increased asymmetry and trunk oscillations (Liao et al., 2015; Li et al., 2018). The innervation of the trunk muscles is provided both as a contralateral dominant and ipsilateral hemisphere, so post-stroke involvement is seen on both sides (Onursal, 2015). Based on this view, in the study comparing the effects of routine physiotherapy with bilateral upper extremity exercises, it was reported that the upper extremity functions and daily life activity performances of the study group, which included bilateral exercise program in hemiplegic patients, improved more than the control group who was rehabilitated only hemiplegic side (Lee et al., 2017). In another study that evaluating the grip strength after stroke, it was found that the unaffected side grip strength of the hemiplegic group was significantly lower than that of healthy controls, and the researchers stated that the unaffected side should be included in the treatment programs (Park & Park, 2016).

The central nervous system provides preliminary postures (APAs) and compensatory posture arrangements. These arrangements appear at the beginning of the extremity movements or throughout the movement, provide space orientation, and prepare the body against movements that will disrupt stabilization (Onursal, 2015). EMG studies indicated that hemiplegic patients had delayed APAs in both hemiplegic and non-hemiplegic sides. This delay is primarily in the M. Erector Spina, M. Latissimus Dorsi, and M. External Oblique; Besides, it can be seen in other muscles according to the character of the movement (Lee et al., 2018). In a study, it was stated that the APAs delay varies according to the character of the movement and it is suggested to use dual-task activities to obtain positive results for the function (Hwang et al., 2015).

In stroke patients, the hemiplegic side trunk is affected negatively. The non-hemiplegic side has to support the hemiplegic side and close its deficiency during functional movements. The patient carries out the activity with various compensations. This results in the non-hemiplegic side having a different structural and functional feature than healthy individuals over time. There are many reasons for trunk impairment after stroke such as muscle weakness, balance problems, postural impairments. Trunk rehabilitation improves trunk's and trunk associated functions (Büyükavcı et al., 2016; Souzaa et al., 2019; Zhang et al., 2020; Sharma & Kaur, 2017; Shumway-Cook & Woollacott, 2016; Granacher et al., 2013; Liu et al., 2020).

## ASSESSMENT METHODS OF THE TRUNK

In stroke rehabilitation, it is indispensable the evaluation of trunk movements to determine the needs of the patient and to plan the treatment program, and to increase the quality of daily life, which is the main goal of the rehabilitation. There are many evaluation methods, devices, and scales for trunk function and performance (Wade et al., 1983; Sandin & Smith, 1990; Bohannon, 1992).

### 1. Trunk Assessment Scales

#### a. The Motor Assessment Scale-trunk subscale

The scale developed by Carr et al. consists of items that evaluate trunk performance after stroke. Each item is scored on a scale of 0 to 6. The highest score indicates the best performance (Balkan et al., 2019).

#### b. The Stroke Impairment Assessment Set- Trunk Control Subscale

This scale evaluates the trunk in two ways. One of them is vertical and the other is an abdominal manual muscle test, both of which are evaluated as four-point sequential scales. The highest score indicates the best performance (Tsuji et al., 2007).

#### c. Trunk Control Test

The Trunk Control Test is a special clinical test that evaluates the motor performance of the trunk. The test consists of four items and is an ordered scale of 3 points. Substances:

i. Rolling over to the strong and weak sides in a supine position,
ii. Coming to sit from a lying position,
iii. The balance is evaluated in the sitting position on the edge of the bed.

The total score on this test ranges from a minimum of 0 to a maximum of 100 points, a higher score means better performance (0- Cannot move without help; 12- Performs the movement in an abnormal manner;25- Complements the movement normally) (Collin & Wade, 1990).

In this test, the patient's turning to the weak and strong side, lying in a sitting position and sitting on the edge of the bed in a balanced position with his feet high from the ground for a minimum of 30 seconds, being able to hold himself while sitting and standing at a height of 30 seconds are evaluated (Wade & Hewer, 1987).

### d. Trunk Impairment Scale

This scale is used to evaluate the motor impairment of the trunk after stroke. This scale consists of 17 items; evaluates the static, dynamic sitting balance, and trunk coordination. The total score gets a minimum of 0 to a maximum of 23 points. A high score indicates good trunk performance. It can control the patient's healing process. In this scale;

i. In static settlement balance assessment; evaluates whether the person can sit independently with their legs crossed.
ii. In dynamic sitting balance; evaluates the ability to actively contract both sides of the trunk, initiated from the shoulder girdle or pelvic belt.
iii. Trunk coordination evaluates the ability to rotate the shoulder girdle and pelvic girdle (Verheyden et al., 2004).
iv. Trunk Impairment Scale measures perception of trunk verticality on the affected and unaffected side, trunk rotation muscle strength, and correction of reflexes on the affected and unaffected side (Fujiwara et al., 2004).

## 2. Trunk Assessment Devices

Trunk stabilization; gravity is the ability to maintain control over the posture and movements of the body against internal and external stimuli. Trunk stabilization is mechanically linked to passive (bone structure), active (muscle structure), and neural system (Cholewicki & McGill, 1996).

Stabilization of the trunk; necessary to control trunk movement during daily activities such as standing, sitting, walking (van der Burg et al., 2005). The affection of the trunk restricts arm and hand mobility and negatively affects daily life activities. (Pigeon et al., 2000). There are limited devices that measure the trunk function directly. So we included devices that evaluate the trunk function indirectly.

## a. Accelerometer

The accelerometer gives information about balance, and gait and indirectly about the trunk. Although its clinical use is not very common in stroke patients, trunk accelerometer can be used in walking and steps. (Osaka et al., 2017). The accelerometer is fixed to the L3 level with velcro and the walking cycle is recorded (Iosa et al., 2012). Vertical, anterior-posterior, and mediolateral measurements of the trunk are made with trunk accelerometer. Step regularity is determined by the autocorrelation coefficient of trunk acceleration in the gait cycle during walking (Bautmans et al., 2011).

## b. Isokinetic Dynamometer

Isokinetic dynamometers combined with a body module moving in the sagittal plane are used to evaluate the trunk torque. Eccentric, isometric, and concentric peak torque of the trunk flexor and extensor muscles may be measured by the dynamometer (Maffiluetti et al., 2007).

## c. Hand-Held Dynamometer

This dynamometer is used to evaluate the strength of the trunk muscles indirectly. The hand-held dynamometer has intra and inter-tester reliability to quantify the trunk muscle strength in patients with chronic stroke (Bohannon, 1992; Burns & Spanier, 2005; Karthikbabu & Chakrapani, 2017).

## d. Electromyography (EMG)

EMG activity of the trunk muscles give us information about performance of the trunk during the activity. It was determined that the trunk muscles were insufficient in the postural control in the anterior-posterior, axial-lateral directions while performing the flexion movements of the upper and lower extremities after stroke (Karthikbabu et al., 2012; Dickstein et al., 2004). The erector spinae and rectus abdominis muscles are dynamically evaluated in sitting and standing up with EMG in patients with stroke. (Dickstein et al., 1999). Placement of electrodes;

i. For rectus abdominis muscles; It is placed bilaterally over the rectus abdominis muscles, 3 cm parallel to the muscle fibers and 2 cm lateral to the umbilicus.
ii. For erector spinae; The electrodes are placed 2 cm apart and 2 cm lateral to either side of the L3 vertebra. Placed parallel to the vertebral column (Cram et al., 1998).

### e. 3D Kinematic

3D Kinematic measures the dynamic and static balance of the trunk in the sitting position. This evaluation uses 8 cameras at 340 Hz. markers; C7 is placed in 6 anatomical points: right and left shoulder, S1, right and left spina iliaca posterior. In stroke patients, the sitting position is dynamically evaluated by repeating movements in the anterior, posterior and lateral directions for 5 times. The patient rests for 5 seconds between each task (Carozzoa et al., 2020).

## REHABILITATION METHODS OF THE TRUNK

One of the aims of stroke rehabilitation is to improve the problems that may affect the quality of daily life. Most studies focus on upper and lower extremity rehabilitation to improve performance and neglected the trunk. But the holistic approach is the best method for stroke rehabilitation (Van Peppen et al., 2004; Verheyden et al., 2007). It is stated that with loss of trunk control, respiratory, speech, balance, walking, arm, and hand functions are affected (Davies, 1990).
There are different rehabilitation methods of stroke in the literature.

- Computer-aided balance exercises (Nintendo Wii Fit-heading, table tilt, balance bubble.
- Proprioceptive Neuromuscular Facilitation Exercises
- Core stability exercises
- Trunk exercises plus Neurodevelopmental Therapy
- Task-specific trunk exercises
- Matrix ritm therapy
- Sling Exercises Therapy (Büyükavcı et al., 2016; Souzaa et al., 2019; Zhang et al., 2020; Sharma & Kaur, 2017; Shumway-Cook & Woollacott, 2016; Granacher et al., 2013; Liu et al., 2020).

## CONCLUSION

There are many different assessment and rehabilitation methods for the trunk. In humans, the trunk is the key point of the function. Trunk evaluation and rehabilitation in stroke patients should not be neglected. There are many methods to be used in trunk rehabilitation. The right method should be applied to the right patient by choosing the most appropriate method for the patient.

## REFERENCES

Balkan, F. A., Salcı, Y., Keklicek, H., Çetin, B., Adın, R. M., & Armutlu, K. (2019). The trunk control: Which scale is the best in very acute stroke patients? *Topics in Stroke Rehabilitation*, *26*(5), 359–365. doi:10.1080/10749357.2019.1607994 PMID:31044664

Bautmans, I., Jansen, B., Van Keymolen, B., & Mets, T. (2011). Reliability and clinical correlates of 3D-accelerometry based gait analysis outcomes according to age and fall-risk. *Gait & Posture*, *33*(3), 366–372. doi:10.1016/j.gaitpost.2010.12.003 PMID:21227697

Bohannon, R. W. (1992). Lateral trunk flexion strength: Impairment, measurement reliability and implications following unilateral brain lesion. *International Journal of Rehabilitation Research. Internationale Zeitschrift fur Rehabilitationsforschung. Revue Internationale de Recherches de Readaptation*, *15*(3), 249–251. doi:10.1097/00004356-199209000-00010 PMID:1428392

Bohannon, R. W., Cassidy, D., & Walsh, S. (1995). Trunk muscle strength is impaired multidirectionally after stroke. *Clinical Rehabilitation*, *9*(1), 47–51. doi:10.1177/026921559500900107

Burns, S. P., & Spanier, D. E. (2005). Break- technique hand held dynamometry: Relation between angular velocity and strength measurements. *Archives of Physical Medicine and Rehabilitation*, *86*(7), 1420–1426. doi:10.1016/j.apmr.2004.12.041 PMID:16003675

Büyükavcı, R., Şahin, F., Sağ, S., Doğu, B., & Kuran, B. (2016). The impact of additional trunk balance exercises on balance, functional condition and ambulation in early stroke patients: Randomized controlled trial. *Turkish Journal of Physical Medicine and Rehabilitation*, *62*(3), 248–256. doi:10.5606/tftrd.2016.84770

Carozzoa, S., Serraa, S., Pignoloa, L., Tonina, P., & Cerasa, A. (2020). The assessment of trunk recovery in stroke patients using 3D kinematic measures. *Medical Engineering & Physics*, *78*, 98–105. doi:10.1016/j.medengphy.2020.01.013 PMID:32035812

Cholewicki, J., & McGill, S. M. (1996). Mechanical stability of the in vivo lumbar spine: Implications for injury and chronic low back pain. *Clinical Biomechanics (Bristol, Avon)*, *11*(1), 1–15. doi:10.1016/0268-0033(95)00035-6 PMID:11415593

Collin, C., & Wade, D. (1990). Assessing motor impairment after stroke: A pilot reliability study. *Journal of Neurology, Neurosurgery, and Psychiatry*, *53*(7), 576–579. doi:10.1136/jnnp.53.7.576 PMID:2391521

Cram, J. R., Kasman, G. S., & Holtz, J. (1998). *Introduction to Surface Electromyography*. Aspen Publishers Inc.

Davies, P. M. (1990). Right in the middle. Springer-Verlag. doi:10.1007/978-3-642-61502-3

Dickstein, R., Heffes, Y., Laufer, Y., & Ben-Haim, Z. (1999). Activation of selected trunk muscles during symmetric functional activities in poststrokehemiparetic and hemiplegic patients. *Journal of Neurology, Neurosurgery, and Psychiatry*, *66*(2), 218–221. doi:10.1136/jnnp.66.2.218 PMID:10071103

Dickstein, R., Shefi, S., Marcovitz, E., & Villa, Y. (2004). Anticipatory postural adjustment in selected trunk muscles in post stroke hemiparetic patients. *Archives of Physical Medicine and Rehabilitation*, *85*(2), 261–267. doi:10.1016/j.apmr.2003.05.011 PMID:14966711

Fujiwara, T., Liu, M., Tsuji, T., Sonoda, S., Mizuno, K., Akaboshi, K., Hase, K., Masakado, Y., & Chino, N. (2004). Development of a new measure to assess trunk impairment after stroke (trunk impairment scale): Its psychometric properties. *American Journal of Physical Medicine & Rehabilitation*, *83*(9), 681–688. doi:10.1097/01.PHM.0000137308.10562.20 PMID:15314532

Granacher, U., Lacroix, A., Muehlbauer, T., Roettger, K., & Gollhofer, A. (2013). Effects of core instability strength training on trunk muscle strength, spinal mobility, dynamic balance and functional mobility in older adults. *Gerontology*, *59*(2), 105–113. doi:10.1159/000343152 PMID:23108436

Hwang, W. J., Cho, M. K., & Chung, Y. (2015). Relationship between anticipatory postural adjustment of the trunk, dual tasks and physical performance with chronic stroke survivors: A pilot test. *Physical Therapy Rehabilitation Science*, *4*(1), 44–48. doi:10.14474/ptrs.2015.4.1.44

Iosa, M., Fusco, A., Morone, G., Pratesi, L., Coiro, P., Venturiero, V., De Angelis, D., Bragoni, M., & Paolucci, S. (2012). Assessment of upper-body dynamic stability during walking in patients with subacute stroke. *Journal of Rehabilitation Research and Development*, *49*(3), 439–450. doi:10.1682/JRRD.2011.03.0057 PMID:22773202

Karthikbabu, S., & Chakrapani, M. (2017). Hand-Held Dynamometer is a Reliable Tool to Measure Trunk Muscle Strength in Chronic Stroke. *Journal of Clinical and Diagnostic Research: JCDR*, *11*(9), YC09–YC12. doi:10.7860/JCDR/2017/28105.10672 PMID:29207821

Karthikbabu, S., Chakrapani, M., Ganeshan, S., Rakshith, K. C., Nafeez, S., & Prem, V. (2012). A review on assessment and treatment of the trunk in stroke: A need or luxury. *Neural Regeneration Research*, *7*(25), 1974–1977. PMID:25624827

Lee, M. J., Lee, J. H., Koo, H. M., & Lee, S. M. (2017). Effectiveness of bilateral arm training for improving extremity function and activities of daily living performance in hemiplegic patients. *Journal of Stroke and Cerebrovascular Diseases*, *26*(5), 1020–1025. doi:10.1016/j.jstrokecerebrovasdis.2016.12.008 PMID:28162905

Lee, N. G., You, J. S. H., Chung, H. Y., Jeon, H. S., Choi, B. S., Lee, D. R., & Yoon, H. S. (2018). Best Core Stabilization for Anticipatory Postural Adjustment and Falls in Hemiparetic Stroke. *Archives of Physical Medicine and Rehabilitation*, *99*(11), 2168–2174. doi:10.1016/j.apmr.2018.01.027 PMID:29476713

Li, S., Francisco, G. E., & Zhou, P. (2018). Post-stroke hemiplegic gait: New perspective and insights. *Frontiers in Physiology*, *9*, 1021. doi:10.3389/fphys.2018.01021 PMID:30127749

Liao, C. F., Liaw, L. J., Wang, R. Y., Su, F. C., & Hsu, A. T. (2015). Electromyography of symmetrical trunk movements and trunk position sense in chronic stroke patients. *Journal of Physical Therapy Science*, *27*(9), 2675–2681. doi:10.1589/jpts.27.2675 PMID:26504267

Liu, J., Feng, W., Zhou, J., Huang, F., Long, L., Wang, Y., Liu, P., Huang, X., Yang, M., Wang, K., & Sun, Z. (2020). Effects of sling exercise therapy on balance, mobility, activities of daily living, quality of life and shoulder pain in stroke patients: A randomized controlled trial. *European Journal of Integrative Medicine*, *35*(5), 1010772. doi:10.1016/j.eujim.2020.101077

Maffiluetti, N. A., Bizzini, M., Desbrosses, K., Babault, N., & Munzinger, U. (2007). Relialibilty of knee extension and flexion measurements using the Con-Trex isokinetic dynamometer. *Clinical Physiology and Functional Imaging*, *27*(6), 346–353. doi:10.1111/j.1475-097X.2007.00758.x PMID:17944656

Onursal, F. Ö. (2015). *Inmede Gövde Eğitimi. Fizyoterapistler ve öğrenciler için e-kitap Bahar Dönemi*. https://www.researchgate.net/publication/329717723

Osaka, H., Shinkoda, K., Watanabe, S., Fujita, D., Kobara, K., Yoshimura, Y., Ito, T., & Suehiro, T. (2017). Association between trunk acceleration during walking and clinically assessed balance in patients with stroke. *NeuroRehabilitation*, *41*(4), 783–790. doi:10.3233/NRE-172171 PMID:29254113

Park, S., & Park, J. Y. (2016). Grip strength in post-stroke hemiplegia. *Journal of Physical Therapy Science*, *28*(2), 677–679. doi:10.1589/jpts.28.677 PMID:27065562

Pigeon, P., Yahia, L. H., Mitnitski, A. B., & Feldman, A. G. (2000). Superposition of independent units of coordination during pointing movements involving the trunk with and without visual feedback. *Experimental Brain Research, 131*(3), 336–349. doi:10.100700221900267 PMID:10789948

Sandin, K. J., & Smith, B. S. (1990). The measure of balance in sitting in stroke rehabilitation prognosis. *Stroke, 21*(1), 82–86. doi:10.1161/01.STR.21.1.82 PMID:2300995

Sharma, V., & Kaur, J. (2017). Effect of core strengthening with pelvic proprioceptive neuromuscular facilitation on trunk, balance, gait, and function in chronic stroke. *Journal of Exercise Rehabilitation, 13*(2), 200–205. doi:10.12965/jer.1734892.446 PMID:28503533

Shumway-Cook, A., & Woollacott, M. H. (2016). Motor Control: Translating Research Into *Clinics and Practice*. Wolters Kluwer.

Souzaa, D. C. B., Santosb, M. S., Ribeirob, N. M. S., & Maldonado, I. L. (2019). Inpatient trunk exercises after recent stroke: An update meta-analysis of randomized controlled trials. *NeuroRehabilitation, 44*(3), 369–377. doi:10.3233/NRE-182585

Tanaka, S., Hachisuka, K., & Ogata, H. (1997). Trunk rotatory muscle performance in post-stroke hemiplegic patients1. *American Journal of Physical Medicine & Rehabilitation, 76*(5), 366–369. doi:10.1097/00002060-199709000-00003 PMID:9354489

Tanaka, S., Hachisuka, K., & Ogata, H. (1998). Muscle strength of trunk flexion-extension in post-stroke hemiplegic patients. *American Journal of Physical Medicine & Rehabilitation, 77*(4), 288–290. doi:10.1097/00002060-199807000-00005 PMID:9715916

Tsuji, T., Liu, M., Sonoda, S., Domen, K., & Chino, N. (2000). The stroke impairment assessment set: Its internal consistency and predictive validity. *Archives of Physical Medicine and Rehabilitation, 81*(7), 863–868. doi:10.1053/apmr.2000.6275 PMID:10895996

Van der Burg, J. C., Pijnappels, M., & van Dieen, J. H. (2005). Out of plane trunk movements and trunk muscle activity after a trip during walking. *Experimental Brain Research, 165*(3), 407–412. doi:10.100700221-005-2312-z PMID:15875169

Van Peppen, R. P., Kwakkel, G., Wood-Dauphinee, S., Hendriks, H. J., Van der Wees, P. J., & Dekker, J. (2004). The impact of physical therapy on functional outcomes after stroke: What's the evidence? *Clinical Rehabilitation, 18*(8), 833–862. doi:10.1191/0269215504cr843oa PMID:15609840

VanCriekinge, T., Saeys, W., Hallemans, A., Velghe, S., Viskens, P. J., Vereeck, L., De Hertogh, W., & Truijen, S. (2017). Trunk biomechanics during hemiplegic gait after stroke: A systematic review. *Gait & Posture*, *54*, 133–143. doi:10.1016/j.gaitpost.2017.03.004 PMID:28288334

Verheyden, G., Nieuwboer, A., Mertin, J., Preger, R., Kiekens, C., & De Weerdt, W. (2004). The trunk impairment scale: A new tool to measure motor impairment of the trunk after stroke. *Clinical Rehabilitation*, *18*(3), 326–334. doi:10.1191/0269215504cr733oa PMID:15137564

Verheyden, G., Nieuwboer, A., Van de Winckel, A., & De Weerdt, W. (2007). Clinical tools to measure trunk performance after stroke: A systematic review of the literatüre. *Clinical Rehabilitation*, *21*(5), 387–394. doi:10.1177/0269215507074055 PMID:17613559

Wade, D. T., & Hewer, R. L. (1987). Motor loss and swallowing difficulty after stroke: Frequency, recovery and prognosis. *Acta Neurologica Scandinavica*, *76*(1), 50–54. doi:10.1111/j.1600-0404.1987.tb03543.x PMID:3630644

Wade, D. T., Skilbeck, C. E., & Hewer, R. L. (1983). Predicting Barthel ADL score at 6 months after an acute stroke. *Archives of Physical Medicine and Rehabilitation*, *64*(1), 24–28. PMID:6849630

Zhang, Q., Fu, C., Liang, Z., Peng, L., Xiong, F., Chen, L., He, C., & Wei, Q. (2020). The effect of adding trunk restraint to task-oriented training in improving function in stroke patients: A systematic review and meta-analysis. *NeuroRehabilitation*, *46*(1), 95–108. doi:10.3233/NRE-192964 PMID:32039877

# Chapter 12
# Vestibular Rehabilitation:
## Anatomy and Physiology of the Vestibular System

**Elif Tuğba Sarac**
*Hatay Mustafa Kemal University, Turkey*

## ABSTRACT

*Vestibular rehabilitation (VR) is a therapeutic approach prepared specifically for each individual who has a vestibular and balance disorder. VR helps in the treatment of unilateral or bilateral vestibular hypofunction and vestibular problems such as labyrinthitis and vestibular neuronitis. Individuals who have inner ear problems which have not been solved for a long time or have received medical treatment benefit from VR. In addition, VR helps to alleviate the complaints of individuals who have undergone surgery due to vestibular problems. With the VR program, regulative activities are carried out to decrease the duration, intensity, and frequency of vertigo; the symptoms of vertigo; increase independency in daily life activities; and to make it possible for patients to deal with the feelings of dizziness, imbalance, and anxiety, in addition to training patients about this issue and regulating the general conditions. The aim is to increase the life quality of patients.*

DOI: 10.4018/978-1-7998-7433-1.ch012

## INTRODUCTION

The central and peripheral parts of the vestibular system can be analyzed. The peripheral vestibular system is also known as the vestibular end organ.

### The Peripheral Vestibular System

The vestibular labyrinth is located within the inner ear and consists of the utricle, saccule, and semicircular canals. There are three semicircular canals. The posterior semicircular canal creates a 45° angle in the sagittal and coronal planes. The horizontal (lateral) semicircular canal creates a 30° angle in the horizontal plane. The other semicircular canal is the anterior semicircular canal. The dilated structures at the end of the semicircular canals are called the ampulla. There are crista and cupulas inside the ampulla. The utricle and saccule are also known as static macula. When the position of the head is straight, the utricle is in a horizontal position and the saccule is in a vertical position. Therefore, the utricle perceives the horizontal position of the head, and the saccule perceives the vertical position of the head.

### The Central Vestibular System

The vestibular nerve contains the vestibular nuclei in the brain stem. Since the vestibular nuclei processes other data that are also vestibular, it affects the functions of the oculomotor and spinal motor systems. The vestibular nuclei are connected through visual, spinal, and cerebellar afferents. Visual afferents—data obtained with the movement of visual objects on the retina—are transmitted to the vestibular nuclei. This explains how visual stimulation causes vertigo. For spinal afferents, there are fibers in the joint and muscle receptors of the neck. These fibers transmit information about the positions of the head and the eyes to the vestibular nuclei. With cerebellar afferents, the cerebellum functions as a control center (Bauer & Girardi, 2002).

### The Physiology of the Vestibular System

The vestibular system has three basic functions: 1) to prevent falling when standing up; 2) to make possible the perception of the direction and speed of the position and movements of the head through space with the central nervous system; and 3), to control eye movement when a person moves or objects move around that person.

To summarize, the visual coordinates muscle movements that emerge due to proprioceptive and vestibular inputs in the central nervous system. The peripheral vestibular system allows the direction of gravity to be perceived, which is important to the preservation of balance and to distinguishing the direction in which the head moves. The angular acceleration of the head transmits data about the head and gravity to the central nervous system or the vestibular nuclei. Visual inputs provide the most reliable information that the central nervous system needs to deliver balance to a mobile organism. The necessary information is transmitted through the retinal receptors and n. opticus to the visual cortex in the occipital lobe. This allows voluntary and involuntary eye movements to be regulated.

Proprioceptors are also critical to input. The central nervous system provides information about balance when an organism changes position and data about the movements and locations of the extremities. The receptors in the feet and knees, muscles which work against gravity, are specialized to transmit information about the position of the skin on the soles of the feet. The proprioceptive system enables organisms to maintain balance with the inputs they receive from the skin, joint tendons, and muscles. Vestibular inputs also transmit data about balance to the central nervous system. These inputs inform the central nervous system about the movements of the head through the vestibular labyrinth. Information obtained from these three inputs is accumulated in the vestibular nuclei in the brainstem. A comparison is made about the right and left data, and these are sent to the related centers in the central nervous system (Ceryan & MB, 2007).

## VESTIBULAR REHABILITATION

The data collected from our sensory systems gradually increase from birth and gather in a center that is derived out of reticular formation. This system is under the continuous control of visual, proprioceptive, and vestibular stimulation. It is also impacted by various factors, such as movement, learning, and change. A constant comparison of the stored senses takes place at the subconscious level, and the relevant muscle groups are stimulated to maintain balance. If the information stored in the center comes across differently out of ordinary movement, it is immediately detected. The unwanted reactions continue until information about a new sense is created by the center. The creation of these new models is constant throughout daily life and is known as adaptation(Glasscock, Cueva, & Thedinger, 1990).

Imbalance emerges as vertigo and dizziness and is caused by the failure to achieve body posture or the inability to move appropriately. The terms vertigo and dizziness are generally compared with and used as substitutes for one another. **Vertigo** is the deterioration in the relationship between an individual and his or her environment, wherein the individual feels as if they are falling, spinning, or being pushed. **Dizziness** is the feeling of doziness or being in Space(Dix, 1974; HALMAGYI & Cremer, 2000; Sharpe & Barber, 1993; Yılmaz, 2004).

When the sensory systems provide insufficient or incorrect information, the errors in the central interpretation of these data causes an individual to feel vertigo. In some cases, reinterpreting the data in a conscious manner can improve this state (Glasscock et al., 1990). In cases of long-term vertigo or dizziness where compensation is not developed, vestibular rehabilitation (VR) that is unique to the individual should be considered.

VR is a therapeutic approach that is prepared specifically for individuals who have vestibular or balance disorders. VR helps in the treatment of unilateral or bilateral vestibular hypofunction and vestibular problems, such as labyrinthitis and vestibular neuronitis. VR has been proven to benefit individuals who have received medical treatment with unsolved inner ear problems. In addition, the technique can help alleviate the complaints of individuals who have undergone surgery due to vestibular problems. With the VR program, regulative activities are carried out to decrease the duration, intensity, frequency, and symptoms of vertigo, increase independence in daily life activities, help patients cope with feelings of dizziness, imbalance, and anxiety, and train patients to understand and regulate the general conditions of these issues. Although VR is intended to improve patients' quality of life(TEKİN DAL, 2019), there are certain circumstances, such as cerebellar degeneration and idiopathic movement intolerance, where its effectiveness remains unclear(Yılmaz, 2004).

It is difficult to treat patients whose imbalance persists after they have undergone numerous types of medical or surgical treatment. Movements that require repeated positions and posture changes can cause an individual to develop disabilities at an early age. In addition, balance problems that come with old age can limit a person's quality of life and further incapacitate them due to the secondary risk of injury (Keim, Cook, & Martini, 1992; TEKİN DAL, 2019).

Trained therapists plan VR to help patients compensate with a loss of balance to their vestibular systems. VR consists of balance activities and eye exercises. Balance activities are designed to help patients use their senses in their feet, angles of vision, and remaining vestibular functions so that they can maintain balance. Eye exercises help the brain to learn and reprogram a person's movements. VR programs should be planned and tailored to treat specific individuals(Teichroew, 2016).

VR is designed to train patients, correct balance, and improve symptoms, such as imbalance, by decreasing disability and repairing impaired conditions. The treatment programs that involve these exercises can be analyzed under the following three categories: 1) training exercises; 2) exercises to retrain balance; and 3), efforts to improve general conditions(Shepard, Smith-Wheelock, Telian, & Raj, 1993).

**Training exercises** functionally stimulate and repeat the actions that spur physiological exhaustion due to the involuntary reactions of movement. Patients who have a few symptoms or symptoms that do not reappear spontaneously respond poorly to these exercises(Teichroew, 2016; Telian, 1991). The exercises are selected to include the movements and positions that created symptoms during evaluation. They are then carried out quickly to stimulate the patients' symptoms. It is sufficient to repeat these exercises twice each day; excess exercise can exhaust the compensation system and increase symptoms. The number of exercises is increased gradually. The compensation effect appears as early as the second week(Teichroew, 2016; Telian, 1991).

**Exercises to retrain balance** involve activities to improve disrupted balance.

During the initial steps of rehabilitation, patients should be informed that they will gradually be able to return the normal lives. It should also be emphasized that continuity, work, and performing physical activities on their own will be critical to overcoming their exhaustion and insufficiencies. The activities should be repeated regularly every day(Bauer & Girardi, 2002; Dix, 1974; Jacobson, Newman, & Kartush, 1993; Shepard et al., 1993). The Cawthorne-Cooksey exercises are the most frequently used techniques for reteaching balance. These exercises can help to decrease vertigo by stimulating the vestibular system or by making the vestibular system work(Keim et al., 1992; Shepard et al., 1993; Yılmaz, 2004).

**Efforts to improve general conditions** include activities and areas of engagement that are introduced to patients by their therapists. Patients are taught how to concentrate and to resume the normal activities they performed before they required rehabilitation. These instructions help patients to become self-sufficient and reacquire their previous life conditions. Decisions are made about the safety of the patients' homes and the subsidiary tools that need to be used. These efforts enable patients to achieve full independence and preserve their self-confidence and independence (Dix, 1974; Keim et al., 1992; Telian, 1991).

## DIRECTIVES TO PREVENT FALLING

In order to help patients preserve balance and prevent injuries that may take place due to falling, directives to prevent falling should be applied. These directives should be implemented for all individuals with imbalance, regardless if they have been tested positive for a vestibular disorder (Teichroew, 2016).

1. Always use the walking device provided to you (crutches, walking sticks, etc.).
2. Move slow enough to prevent imbalance when changing positions.
3. Use the elevator instead of stairs or ramps and use railings.
4. Do not use very soft and low chairs and slide to the edge of the chair before getting up.
5. Wear shoes that do not slide with laces and low heels.

6. When standing up from a lying position, sit for a few minutes before slowly getting up. Wait for a few moments and then walk.
7. Changes in the home:
    a. Keep your lights on in the dark and make sure that the light switches are at an easily reachable distance.
    b. Keep cables and small objects away from your walking path.
    c. Keep your phone with you at all times.
    d. Use wall to wall carpeting.
    e. Place the most frequently used objects on easily accessible shelves.
    f. Use mats that do not slip in the bathroom and shower. Use a shower stool. Avoid bathtubs.
    g. If necessary, have bars installed in your bathroom and shower area.
8. Do not drive if your quick head movements result in imbalance (Teichroew, 2016).

## CONCLUSION

Basically, the visual coordinates the muscle movements which emerge as a result of the process of the proprioceptive and vestibular inputs in the central nervous system. The duty of the peripheral vestibular system is to perceive which direction gravity which is important in the preservation of balance goes and which direction the head moves towards. The angular acceleration of the head transmits data to the central nervous system or the vestibular nuclei about the orientation of the head in terms of gravity. The visual inputs provide the most reliable and needed information to the central nervous system in order to provide the balance of an organism which is mobile. The required information through the retinal receptors and n. opticus are transmitted to the visual cortex located in the occipital lobe. Thus, the voluntary and involuntary eye movements are regulated. Another important input is the proprioceptors. The central nervous system aims at providing balance in line which the changing position and giving information about the movements and location of the extremities. The receptors in the feet and the knees, muscles which work against gravity are specialized to transmit information about the position of the skin of the soles of the feet. The proprioceptive system allows keeping the organism in balance with the inputs it receives from the skin, joint tendons and muscles. The other input which transmits data to the central nervous system about balance is the vestibular inputs. These inform the central nervous system through the vestibular labyrinth in line with the movements of the head. The information obtained from these three inputs is accumulated in the vestibular nuclei in the brainstem. A comparison is made about the right and left data and these are sent to the related centers in the central nervous system(Ceryan & MB, 2007).

## REFERENCES


Bauer, C., & Girardi, M. (2002). Vestibular rehabilitation. *eMedicine Journal, 3*(6).

Ceryan, K., & MB, Ş. (2007). Baş dönmesi olan hastada öykü, muayene ve tanı. *Kulak burun boğaz ve baş boyun cerrahisi, 2*, 36-63.

Dix, M. (1974). Treatment of vertigo. *Physiotherapy, 60*(12), 380–384. PMID:4467199

Glasscock, M. E., Cueva, R. A., & Thedinger, B. A. (1990). *Handbook of vertigo*. Raven Press.

Halmagyi, G. M., & Cremer, P. (2000). Assessment and treatment of dizziness. BMJ Publishing Group Ltd.

Jacobson, G. P., Newman, C. W., & Kartush, J. M. (1993). *Handbook of balance function testing*. Mosby Elsevier Health Science.

Keim, R. J., Cook, M., & Martini, D. (1992). How i do it: Otology and neurotology: A specific issue and its solution: Balance rehabilitation therapy. *The Laryngoscope, 102*(11), 1302–1307. doi:10.1288/00005537-199211000-00019 PMID:1405995

Sharpe, J. A., & Barber, H. O. (1993). *The vestibulo-ocular reflex and vertigo*. Lippincott Williams & Wilkins.

Shepard, N. T., Smith-Wheelock, M., Telian, S. A., & Raj, A. (1993). Vestibular and balance rehabilitation therapy. *The Annals of Otology, Rhinology, and Laryngology, 102*(3), 198–205. doi:10.1177/000348949310200306 PMID:8457121

Teichroew, J. K. (2016). *Chronic Diseases: An Encyclopedia of Causes, Effects, and Treatments* (2 vols.). ABC-CLIO.

Tekin Dal. (2019). Baş Dönmesi Olan Hastaların Vestibüler Rehabilitasyonunda Aktivite Eğitiminin Etkisi. Academic Press.

Telian, S. A. (1991). Physical therapy program for vestibular rehabilitation. *The American Journal of Otology, 12*(3). PMID:1882973

Yılmaz, Ö. (2004). Vestibüler rehabilitasyon. *Ankara Üniveristesi Sağlık Hizmetleri Meslek Yüksekokulu Dergisi, 5*(1), 11–15.

# Glossary

**Diffusion:** Movement of water molecules from an area of higher concentration to lower concentration.

**dMRI:** dMRI stands for diffusion Magnetic Resonance Imaging which is a popular method used by neuroscientists to uncover unique information about the neural connections within the brain. Tractography - A 3D modelling technique used to represent the white matter fibers inside the brain.

**DTI:** DTI refers to diffusion tensor imaging which is a type of deterministic method predominantly used in clinical applications.

**DWI:** DWI refers to diffusion weighted image which is captured with the help of dMRI. Captures the diffusion information of the water molecules inside the brain.

**fODF:** Defined for all fibers in the unit. Indicates the probability of a fiber in a given voxel.

**Voxel:** A 3D unit that embeds the signal in brain scans.

**White Matter Fibers:** Areas of the central nervous system that are made up of myelinated axons. Also referred to as tracts.

# Compilation of References


Dean, J. (2008, January). Toward Enhanced P300 Speller Performance. *Journal of Neuroscience Methods*, *167*(1), 15–21. doi:10.1016/j.jneumeth.2007.07.017 PMID:17822777

Medium. (n.d.). https://medium.com/swlh/graph-theory-machine-learning-in-neuroscience-30f9bec5d182#:~:text=In%20neuroscience%2C%20we%20often%20use,functionally%20connected%20to%20each%20other.&text=We'll%20be%20focusing%20on%20resting%2Dstate%20functional%20connectivity%2C,completing%20any%20kind%20of%20task

Prasolenko, O., Burko, D., Tolmachov, I., Gyulyev, N., Galkin, A., & Lobashov, O. (2019). Creating safer routing for urban freight transportation. *Transportation Research Procedia*, *39*, 417–427. doi:10.1016/j.trpro.2019.06.044

Saposnik, G., Barinagarrementeria, F., Brown, R. D. Jr, Bushnell, C. D., Cucchiara, B., Cushman, M., deVeber, G., Ferro, J. M., & Tsai, F. Y. (2011). Diagnosis and management of cerebral venous thrombosis: A statement for healthcare professionals from the American Heart Association/ American Stroke Association. *Stroke*, *42*(4), 1158–1192. doi:10.1161/STR.0b013e31820a8364 PMID:21293023

Jagušt, T., Botički, I., & So, H. (2018). Examining competitive, collaborative and adaptive gamification in young learners' math learning. *Computers & Education*, *125*, 444–457. doi:10.1016/j.compedu.2018.06.022

Liu, J., Li, M., Pan, Y., & Lan, W. (2017). Complex Brain Network Analysis and Its Applications to Brain Disorders: A Survey. Complexity. doi:10.1155/2017/8362741

Maggi, L., Parini, S., Puccini, L., Panini, G., & Andreoni, G. (2006). A Four Command BCI System Based on the SSVEP Protocol. *IEEE International Conference on Engineering in Medicine and Biology Society*, 1264-1267. 10.1109/IEMBS.2006.260353

Sassi, Touati, Baccouche, Drissi, Romdhane, & Hentati. (2017). Cerebral Venous Thrombosis: A Tunisian Monocenter Study on 160 Patients. *Clinical and Applied Thrombosis/Hemostasis, 23*(8), 1005-9.

Ben, H. (2004, June). An Exploratory Study of Factors Affecting Single Trial P300 Detection. *IEEE Transactions on Biomedical Engineering*, *51*(6), 975–978. doi:10.1109/TBME.2004.826684 PMID:15188867

Ferro, J. M., Canhão, P., Stam, J., Bousser, M. G., & Barinagarrementeria, F. (2004). Prognosis of cerebral vein and dural sinus thrombosis: Results of the International Study on Cerebral Vein and Dural Sinus Thrombosis (ISCVT). *Stroke*, *35*(3), 664–670. doi:10.1161/01.STR.0000117571.76197.26 PMID:14976332

Rubinov, M., & Sporns, O. (2010). Complex network measures of brain connectivity: Uses and interpretations. *NeuroImage*, *52*(3), 1059–1069. doi:10.1016/j.neuroimage.2009.10.003 PMID:19819337

Sharma, A., Khosla, A., & Khosla, M. (2018). Efficacy of an Android-based Game Intervention in Enhancement of Face Recognition Skills for Children with Autism. *International Journal of Learning and Change*, *1*(1), 1. doi:10.1504/IJLC.2018.10011657

Coutinho, J. M., Ferro, J. M., Canhão, P., Barinagarrementeria, F., Bousser, M. G., & Stam, J. (2010). Unfractionated or low-molecular weight heparin for the treatment of cerebral venous thrombosis. *Stroke*, *41*(11), 2575–2580. doi:10.1161/STROKEAHA.110.588822 PMID:20930161

Joyce, K. E. (2010). A new measure of centrality for brain networks. *PloS One, 5*(8).

Kim & Sung-Phil. (2011). Detection of P300 Components using the Wiener Filter for BCI-Based Spellers. *8th Asian IEEE Conference on Control (ASCC),* 892-896.

Naveen Kumar, C., & Shivakumar, G. (2018). *A Real Time Human Emotion Recognition System Using Respiration Parameters and ECG*. Intelligent Human Computer Interaction. doi:10.1007/978-3-030-04021-5_4

Angles, R., & Gutiérrez, C. (2008). Survey of graph database models. *ACM Computing Surveys*, *40*(1), 1–39. doi:10.1145/1322432.1322433

Baumgartner, R. W., Studer, A., Arnold, M., & Georgiadis, D. (2003). Recanalisation of cerebral venous thrombosis. *Journal of Neurology, Neurosurgery, and Psychiatry*, *74*(4), 459–461. doi:10.1136/jnnp.74.4.459 PMID:12640063

Rosas-Cholula. (2010). On Signal P300 Detection for BCI Applications Based on Wavelet Analysis and ICA Preprocessing. *IEEE Conference on Electronics, Robotics and Automotive Mechanics Conference (CERMA), 23*, 360-365.

Udovičić, G., Đerek, J., Russo, M., & Sikora, M. (2017). Wearable Emotion Recognition System based on GSR and PPG Signals. *Proceedings of the 2nd International Workshop on Multimedia for Personal Health and Health Care*. 10.1145/3132635.3132641

Blankertz, B., Müller, K. R., Curio, G., Vaughan, T. M., Schalk, G., Wolpaw, J. R., Schlögl, A., Neuper, C., Pfurtscheller, G., Hinter Berger, T., Schroder, M., & Birbaumer, N. (2004, June). The BCI Competition 2003: Progress and Perspectives in Detection and Discrimination of EEG Single Trials. *IEEE Transactions on Biomedical Engineering*, *51*(6), 1044–1051. doi:10.1109/TBME.2004.826692 PMID:15188876

Ercim. (n.d.). https://ercim-news.ercim.eu/en125/r-i/graph-based-management-of-neuroscience-data-representation-integration-and-analysis

Sharma, R., Goel, D., Srivastav, M., & Dhasmana, R. (2017). Psycho-Physiological Parameters of Nurses in Critical and Non-Critical Units. *International Journal of Nursing Sciences*, *7*(5), 107–110.

Théaudin, M., Crassard, I., Bresson, D., Saliou, G., Favrole, P., Vahedi, K., Denier, C., & Bousser, M.-G. (2010). Should decompressive surgery be performed in malignant cerebral venous thrombosis? A series of 12 patients. *Stroke*, *41*(4), 727–731. doi:10.1161/STROKEAHA.109.572909 PMID:20185780

Cattell, R. (2011). Scalable SQL and NoSQL data stores". *SIGMOD Record*, *39*(4), 12–27. doi:10.1145/1978915.1978919

Cocotte, H., & Graser, A. (2011, March). Convolutional Neural Networks for P300 Detection with Application to BCI,". *IEEE Transactions on Communications*, *33*(3), 433–445.

Gábana Arellano, D., Tokarchuk, L., & Gunes, H. (2016). *Measuring Affective, Physiological and Behavioural Differences in Solo, Competitive and Collaborative Games*. Lecture Notes of the Institute for Computer Sciences, Social Informatics and Telecommunications Engineering. doi:10.1007/978-3-319-49616-0_18

Gulati, D., Strbian, D., & Sundararajan, S. (2014). Cerebral venous thrombosis: Diagnosis and management. *Stroke*, *45*(2), e16–e18. doi:10.1161/STROKEAHA.113.003964 PMID:24357654

Garrett, D., Peterson, D. A., Anderson, C. W., & Thaut, M. H. (2003, June). Comparison of Linear, Nonlinear, and Feature Selection Methods for EEG Signal Classification. *IEEE Transactions on Neural Systems and Rehabilitation Engineering*, *11*(2), 141–144. doi:10.1109/TNSRE.2003.814441 PMID:12899257

Neo4j. (n.d.). https://neo4j.com/blog/data-modeling-basics/

Urh, M., Vukovic, G., Jereb, E., & Pintar, R. (2015). The Model for Introduction of Gamification into E-learning in Higher Education. *Procedia: Social and Behavioral Sciences*, *197*, 388–397. doi:10.1016/j.sbspro.2015.07.154

Carnagey, N., Anderson, C., & Bushman, B. (2007). The effect of video game violence on physiological desensitization to real-life violence. *Journal of Experimental Social Psychology*, *43*(3), 489–496. doi:10.1016/j.jesp.2006.05.003

Kaper, M., Meinicke, P., Grossekathoefer, U., Lingner, T., & Ritter, H. (2004, June). BCI Competition 2003-Dataset IIb: SVM for the P300 Speller Paradigm. *IEEE Transactions on Biomedical Engineering*, *51*(6), 1073–1076. doi:10.1109/TBME.2004.826698 PMID:15188881

Vukotic, A. (2014). *Neo4j in action*. Manning Publications Co.

imotions. (2021). *Galvanic Skin Response (GSR): The Complete Pocket Guide*. Available: https://imotions.com/blog/galvanic-skin-response/

Rakotomamonjy, A., & Gigues, V. (2008, March). BCI competition III: Data Set II-ensemble of SVMs for BCI P300 Speller. *IEEE Transactions on Biomedical Engineering*, *55*(3), 1147–1154. doi:10.1109/TBME.2008.915728 PMID:18334407

Farwell, L. A., & Donchin, E. (1988, December). Talking off the Top of Your Head: Toward a Mental Prosthesis Utilizing Event-Related Brain Potentials. *Electroencephalography and Clinical Neurophysiology*, *70*(6), 510–523. doi:10.1016/0013-4694(88)90149-6 PMID:2461285

Huizenga, W., Admiraal, W., Dam, G., & Voogt, J. (2019). Mobile game-based learning in secondary education: Students' immersion, game activities, team performance and learning outcomes. *Computers in Human Behavior*, *99*, 137–143. doi:10.1016/j.chb.2019.05.020

Masuhr, F., Mehraein, S., & Einhäupl, K. (2004). Cerebral venous and sinus thrombosis. *Journal of Neurology*, *251*(1), 11–23. doi:10.100700415-004-0321-7 PMID:14999484

Teeters, J. L., Harris, K. D., Millman, K. J., Olshausen, B. A., & Sommer, F. T. (2008). Data sharing for computational neuroscience. *Neuroinformatics*, *6*(1), 47–55. doi:10.100712021-008-9009-y PMID:18259695

Lin, Z. L., & Zhang, C. S. (2005). Enhancing Classification by Perceptual Characteristics for the P300 Speller Paradigm. *Proceedings of the 2nd International IEEE EMBS Conference on Neural Engineering*. 10.1109/CNE.2005.1419688

Neulog. (2021). Available: https://neulog.com/Downloads/NeuLog_User_Guide_Ver_6_11b.pdf

Thulasidas, M., Guan, C., & Wu, J. (2006, March). Robust Classification of EEG Signals for BCI. *IEEE Transactions on Neural Systems and Rehabilitation Engineering*, *14*(1), 24–29. doi:10.1109/TNSRE.2005.862695 PMID:16562628

Tobii Pro. (n.d.). Retrieved from https://www.tobiipro.com/learn-and-support/learn/GSR-essentials/how-does-a-gsr- sensor-work/

Matthias Kaper, M. (2004, June). BCI Competition 2003-Data Set IIb: SVM for the P300 Speller Paradigm. *IEEE Transactions on Biomedical Engineering*, *51*(6), 1073–1076.

Maulsby, R., & Edelberg, R. (1960). The interrelationship between the galvanic skin response, basal resistance, and temperature. *Journal of Comparative and Physiological Psychology*, *53*(5), 475–479. doi:10.1037/h0041916 PMID:13768388

Blankertz, B., Curio, G., & Müller, K. R. (2002). Classifying Single Trial EEG: Towards BCI. *Advances in Neural Information Processing Systems*, *14*, 157–164.

ElProCus. (2021). DHT11 Sensor Definition, Working and Applications. *ElProCus - Electronic Projects for Engineering Students*. Available: https://www.elprocus.com/a-brief-on-dht11-sensor/

Schlögl, A., Lee, F., Bischof, H., & Pfurtscheller, G. (2005, June). Characterization of Four-Class Motor Imagery EEG Data for the BCI-competition 2005. *Journal of Neural Engineering*, *2*(4), 14–22. doi:10.1088/1741-2560/2/4/L02 PMID:16317224

Ioana, Nafornita, Boucher, & Isar. (2007). A Bayesian Approach of Hyper Analytic Wavelet Transform Based Denoising. *International Symposium on IEEE, Intelligent Signal Processing*, 1-6.

Sonali, A. (2020). Effect of PUBG on Stress Level in Youngsters Using GSR Sensor. *UGC Care Journal, 40*, 1142–1149.

Brain Signs. (n.d.). *Sapienza University of Rome*. Retrieved from https://www.brainsigns.com/en/science/s2/tehnologies/gsr

John, L. (2004, May). *Semmlow, Biosignal, and Biomedical Image Processing: MATLAB-based Applications*. Academic Press.

Übeyli, E. D. (2009, March). Combined Neural Network Model Employing Wavelet Coefficients for EEG Signals Classification. *Digital Signal Processing, 19*(2), 297–308. doi:10.1016/j.dsp.2008.07.004

Villarejo, M., Zapirain, B., & Zorrilla, A. (2012). A Stress Sensor Based on Galvanic Skin Response (GSR) Controlled by ZigBee. *Sensors (Basel), 12*(5), 6075–6101. doi:10.3390120506075 PMID:22778631

de-Marcos, L., Garcia-Lopez, E., & Garcia-Cabot, A. (2016). On the effectiveness of game-like and social approaches in learning: Comparing educational gaming, gamification & social networking. *Computers & Education, 95*, 99–113. doi:10.1016/j.compedu.2015.12.008

Barlow, D., & Hayes, S. (1979). Alternating treatments design: One strategy for comparing the effects of two treatments in a single subject. *Journal of Applied Behavior Analysis, 12*(2), 199–210. doi:10.1901/jaba.1979.12-199 PMID:489478

Bousser, M. G., & Ferro, J. M. (2007). Cerebral venous thrombosis: An update. *Lancet Neurology, 6*(2), 162–170. doi:10.1016/S1474-4422(07)70029-7 PMID:17239803

Pandey, A. (2021). 6 Killer Examples Of Gamification In eLearning (Updated In 2020). *eLearning Industry*. Available: https://elearningindustry.com/6-killer-examples-gamification-in-elearning

Sporns, O. (2003). Graph theory methods for the analysis of neural connectivity patterns. In *Neuroscience databases* (pp. 171–185). Springer. doi:10.1007/978-1-4615-1079-6_12

Wolpaw, Birbaumer, McFarland, Furtscheller, & Vaughan. (2002). BCI for Communication and Control. *Clinical Neurophysiology, 113*, 767-791.

Abdul Jabbar, A., & Felicia, P. (2015). Gameplay Engagement and Learning in Game-Based Learning. *Review of Educational Research, 85*(4), 740–779. doi:10.3102/0034654315577210

Hwang, G., & Chang, S. (2015). Effects of a peer competition-based mobile learning approach on students' affective domain exhibition in social studies courses. *British Journal of Educational Technology, 47*(6), 1217–1231. doi:10.1111/bjet.12303

Tijs, T., Brokken, D., & IJsselsteijn, W. (2008). Creating an Emotionally Adaptive Game. *Lecture Notes in Computer Science, 5309*, 122–133. doi:10.1007/978-3-540-89222-9_14

Salomon, G. (1984). Television is "easy" and print is "tough": The differential investment of mental effort in learning as a function of perceptions and attributions. *Journal of Educational Psychology*, *76*(4), 647–658. doi:10.1037/0022-0663.76.4.647

Andreassi, J. L. (2007). *Psychophysiology: Human Behavior & Physiological Response* (5th ed.). Psychology Press.

Nozawa, A., & Uchida, M. (2009). *Characterization of preference for viscosity and fragrance of cosmetic emulsions by autonomous nervous system activity*. ICCAS-SICE.

Scotti, S. (2006). Automatic Quantitative Evaluation of Emotions in E-learning Applications. *2006 International Conference of the IEEE Engineering in Medicine and Biology Society*. 10.1109/IEMBS.2006.260601

Wang, W., Weng, J.-F., Su, J.-M., & Tseng, S.-S. (2004). Learning portfolio analysis and mining in SCORM compliant environment. *34th Annual Frontiers in Education. FIE, 2004*, 180–187. Advance online publication. doi:10.1109/FIE.2004.1408490

Villagrasa, S., Fonseca, D., Redondo, E., & Duran, J. (2014). Teaching Case of Gamification and Visual Technologies for Education. *Journal of Cases on Information Technology*, *16*(4), 38–57. doi:10.4018/jcit.2014100104

Werbach, K., & Hunter, D. (2012). For the Win: How game thinking can revolutionize your business. doi:10.1017/CBO9781107415324.004

Agrafioti, F., Hatzinakos, D., & Anderson, A. (2012). ECG Pattern Analysis for Emotion Detection. *IEEE Transactions on Affective Computing*, *3*(1), 102–115. doi:10.1109/T-AFFC.2011.28

De Schutter, E. (2008). Why are computational neuroscience and systems biology so separate? *PLoS Computational Biology*, *4*(5), e1000078. doi:10.1371/journal.pcbi.1000078 PMID:18516226

Dere, F. (2000). *Nöroanatomi Atlası ve Ders Kitabı, üçüncü baskı*. Nobel Yayınları.

Wolpaw, J. R., & McFarland, D. J. (2003, June). The Wadsworth Center BCI Research and Development Program. *IEEE Transactions on Neural Systems and Rehabilitation Engineering*, *11*, 1–4. doi:10.1109/TNSRE.2003.814442

Yildirim, I. (2017). The effects of gamification-based teaching practices on student achievement and students' attitudes toward lessons. *The Internet and Higher Education*, *33*, 86–92. doi:10.1016/j.iheduc.2017.02.002

Arroyo-Gómez, N., Laparra-Hernández, J., Soler-Valero, A., Medina, E., & de Rosario, H. (2017). *Physiological Model to Classify Physical and Cognitive Workload During Gaming Activities*. Advances in Neuroergonomics and Cognitive Engineering. doi:10.1007/978-3-319-60642-2_23

Battiston, F., Nicosia, V., Chavez, M., & Latora, V. (2017). Multilayer motif analysis of brain networks. *Chaos*, *27*(4), 047404. doi:10.1063/1.4979282 PMID:28456158

Mason, S. G., & Birch, G. E. (2003, March). A General Framework for BCI Design. *IEEE Transactions on Neural Systems and Rehabilitation Engineering, 11*, 70–85. doi:10.1109/TNSRE.2003.810426 PMID:12797728

Nagaraja, D., & Sarma, G. R. (2002). Treatment of cerebral sinus/venous thrombosis. *Neurology India, 50*(2), 114–116. PMID:12134170

Ashok Charan. (n.d.). *Marketing analytics*. https://www.ashokcharan.com/Marketing-Analytics/~bm-galvanic-skin-response.php

Cheng, M., Gao, X. R., Gao, S. K., & Xu, D. (2002, March). Design and Implementation of a BCI with High Transfer Rates. *IEEE Transactions on Biomedical Engineering, 49*, 1181–1186. doi:10.1109/TBME.2002.803536 PMID:12374343

Duman, T., Uluduz, D., Midi, I., Bektas, H., Kablan, Y., Goksel, B. K., Milanlioglu, A., Necioglu Orken, D., Aluclu, U., Colakoglu, S., Tüfekci, A., Bakar, M., Nazliel, B., Tascilar, N., Goksan, B., Kozak, H. H., Demir, S., Misirli, C. H., Kucukoglu, H., ... Uzuner, N. (2017). A Multicenter Study of 1144 Patients with Cerebral Venous Thrombosis: The VENOST Study. *Journal of Stroke and Cerebrovascular Diseases, 26*(8), 1848–1857. doi:10.1016/j.jstrokecerebrovasdis.2017.04.020 PMID:28583818

Gao, W., Wu, H., Siddiqui, M. K., & Baig, A. Q. (2018). Study of biological networks using graph theory. *Saudi Journal of Biological Sciences, 25*(6), 1212–1219. doi:10.1016/j.sjbs.2017.11.022 PMID:30174525

Cantu-Brito, C., Arauz, A., Aburto, Y., Barinagarrementeria, F., Ruiz-Sandoval, J. L., & Baizabal-Carvallo, J. F. (2011). Cerebrovascular complications during pregnancy and postpartum: Clinical and prognosis observations in 240 Hispanic women. *European Journal of Neurology, 18*(6), 819–825. doi:10.1111/j.1468-1331.2010.03259.x PMID:21143339

Erdoğan. (2009). *A Design and Implementation of P300 Based BCI* (Ph.D. Thesis). The Graduate School of Natural and Applied Sciences, Middle East Technical University.

Handri, S., Yajima, K., Nomura, S., Ogawa, N., Kurosawa, Y., & Fukumura, Y. (2010). Evaluation of Student's Physiological Response Towards E-Learning Courses Material by Using GSR Sensor. *2010 IEEE/ACIS 9th International Conference on Computer and Information Science*. 10.1109/ICIS.2010.92

Vedavathi & Gurram. (2013). Applications on Graph Theory. *International Journal of Engineering Research & Technology, 2*(1).

Ding, X., Yue, X., Zheng, R., Bi, C., Li, D., & Yao, G. (2019). Classifying major depression patients and healthy controls using EEG, eye tracking and galvanic skin response data. *Journal of Affective Disorders, 251*, 156–161. doi:10.1016/j.jad.2019.03.058 PMID:30925266

Iturria-medina, Y., Sotero, R. C., Canales-rodríguez, E. J., Alemán-gómez, Y., & Melie-garcía, L. (2008). Studying the human brain anatomical network via diffusion-weighted MRI and Graph Theory. *NeuroImage, 40*(3), 1064–1076. doi:10.1016/j.neuroimage.2007.10.060 PMID:18272400

Schaller, B., & Graf, R. (2004). Cerebral venous infarction: The pathophysiological concept. *Cerebrovascular Diseases (Basel, Switzerland), 18*(3), 179–188. doi:10.1159/000079939 PMID:15273432

Zhu, D. H., Berger, J., Molina, G. G., & Arts, R. M. (2010, March). A Survey of Stimulation Methods used in SSVEP-Based BCI system. *Computational Intelligence and Neuroscience, 52*, 702–357. PMID:20224799

Boukobza, M., Crassard, I., & Bousser, M. G. (2007). When the "dense triangle" in dural sinus thrombosis is round. *Neurology, 69*(8), 808. doi:10.1212/01.wnl.0000275548.60964.bf PMID:17709717

Siqueira, S., Eduardo, C., Junior, B., Comfort, W. E., Rohde, L. A., & Sato, J. R. (2014). Abnormal Functional Resting-State Networks in ADHD: Graph Theory and Pattern Recognition Analysis of fMRI Data. *BioMed Research International, 2014*, 380531. PMID:25309910

Towle, L., Bolaños, J., Suarez, D., Tan, K., Grzeszczuk, R., Levin, D. N., Cakmur, R., Frank, S. A., & Spire, J.-P. (1993, January). The Spatial Location of EEG Electrodes: Locating the Best-fitting Sphere Relative to Cortical Anatomy. *Electroencephalography and Clinical Neurophysiology, 86*(1), 1–6. doi:10.1016/0013-4694(93)90061-Y PMID:7678386

Troussas, C., Krouska, A., & Sgouropoulou, C. (2020). Collaboration and fuzzy-modeled personalization for mobile game-based learning in higher education. *Computers & Education, 144*, 103698. doi:10.1016/j.compedu.2019.103698

Farahani, F. V., Karwowski, W., & Lighthall, N. R. (2019). Application of Graph Theory for Identifying Connectivity Patterns in Human Brain Networks: A Systematic Review. *Frontiers in Neuroscience, 13*, 585. doi:10.3389/fnins.2019.00585 PMID:31249501

Nicolas-Alonso, L. F., & Gomez-Gil, J. (2012, January). Brain Computer Interfaces, a Review. *Sensors, 12*(2), 1211–1279. doi:10.3390120201211 PMID:22438708

Tentschert, S., Wimmer, R., Greisenegger, S., Lang, W., & Lalouschek, W. (2005). Headache at stroke onset in 2196 patients with ischemic stroke or transient ischemic attack. *Stroke, 36*(2), e1–e3. doi:10.1161/01.STR.0000151360.03567.2b PMID:15604424

Wang, C., Baird, T., Huang, J., Coutinho, J., Brien, D., & Munoz, D. (2018). Arousal Effects on Pupil Size, Heart Rate, and Skin Conductance in an Emotional Face Task. *Frontiers in Neurology, 9*, 1029. Advance online publication. doi:10.3389/fneur.2018.01029 PMID:30559707

Alhaddi, H. (2015). Triple bottom line and sustainability: A literature review. *Bus. Manag. Stud., 1*(2), 6–10. doi:10.11114/bms.v1i2.752

Al-Qazzaz, N. K., Ali, S. H., Ahmad, S. A., Islam, S., & Mohamad, K. (2014). Cognitive impairment and memory dysfunction after a stroke diagnosis: A post-stroke memory assessment. *Neuropsychiatric Disease and Treatment, 10*, 1677. doi:10.2147/NDT.S67184 PMID:25228808

Amarenco, P. (2001). Cerebellar stroke syndromes. In J. Bogousslavsky & L. R. Caplan (Eds.), *Stroke Syndromes* (pp. 540–556). Cambridge University Press. doi:10.1017/CBO9780511586521.043

Amarenco, P., Albers, G. W., Denison, H., Easton, J. D., Evans, S. R., Held, P., Hill, M. D., Jonasson, J., Kasner, S. E., Ladenvall, P., Minematsu, K., Molina, C. A., Wang, Y., Wong, K. S. L., & Johnston, S. C. (2017). Efficacy and safety of ticagrelor versus aspirin in acute stroke or transient ischaemic attack of atherosclerotic origin: A subgroup analysis of SOCRATES, a randomised, double-blind, controlled trial. *Lancet Neurology*, *16*(4), 301–310. doi:10.1016/S1474-4422(17)30038-8 PMID:28238711

Amarenco, P., Kim, J. S., Labreuche, J., Charles, H., Abtan, J., Béjot, Y., Cabrejo, L., Cha, J.-K., Ducrocq, G., Giroud, M., Guidoux, C., Hobeanu, C., Kim, Y.-J., Lapergue, B., Lavallée, P. C., Lee, B.-C., Lee, K.-B., Leys, D., Mahagne, M.-H., ... Bruckert, E. (2020). A Comparison of Two LDL Cholesterol Targets after Ischemic Stroke. *The New England Journal of Medicine*, *382*(1), 18. doi:10.1056/NEJMoa1910355 PMID:31738483

Amarenco, P., Lavallée, P. C., Labreuche, J., Albers, G. W., Bornstein, N. M., Canhão, P., Caplan, L. R., Donnan, G. A., Ferro, J. M., Hennerici, M. G., Molina, C., Rothwell, P. M., Sissani, L., Školoudík, D., Steg, P. G., Touboul, P.-J., Uchiyama, S., Vicaut, É., & Wong, L. K. (2016). One-Year Risk of Stroke after Transient Ischemic Attack or Minor Stroke. *The New England Journal of Medicine*, *374*(16), 1533–1542. doi:10.1056/NEJMoa1412981 PMID:27096581

Andersen, L. T. (1985). Shoulder pain in hemiplegia. *The American Journal of Occupational Therapy*, *39*(1), 11–19. doi:10.5014/ajot.39.1.11 PMID:2579561

Arboix, A., & Martí-Vilalta, J. L. (2009). Lacunar stroke. *Expert Review of Neurotherapeutics*, *9*(2), 179-196. doi:10.1586/14737175.9.2.179

Arsava, E. M., Bayrlee, A., Vangel, M., Rost, N. S., Rosand, J., Furie, K. L., Sorensen, A. G., & Ay, H. (2011). Severity of leukoaraiosis determines clinical phenotype after brain infarction. *Neurology*, *77*(1), 55–61. doi:10.1212/WNL.0b013e318221ad02 PMID:21700580

Atamaz, F. (2007). Communication disorders and their rehabilitation in stroke. *Turkish Journal of Physical Medicine and Rehabilitation*, *53*(1), 11–15.

Ayerbe, L., Ayis, S., Wolfe, C. D., & Rudd, A. G. (2013). Natural history, predictors and outcomes of depression after stroke: Systematic review and meta-analysis. *The British Journal of Psychiatry*, *202*(1), 14–21. doi:10.1192/bjp.bp.111.107664 PMID:23284148

Ay, H., Koroshetz, W. J., Benner, T., Vangel, M. G., Wu, O., Schwamm, L. H., & Sorensen, A. G. (2005). Transient ischemic attack with infarction: A unique syndrome? *Annals of Neurology*, *57*(5), 679–686. doi:10.1002/ana.20465 PMID:15852402

Azouvi, P., Jacquin-Courtois, S., & Luauté, J. (2017). Rehabilitation of unilateral neglect: Evidence-based medicine. *Annals of Physical and Rehabilitation Medicine*, *60*(3), 191–197. doi:10.1016/j.rehab.2016.10.006 PMID:27986428

Balkan, F. A., Salcı, Y., Keklicek, H., Çetin, B., Adın, R. M., & Armutlu, K. (2019). The trunk control: Which scale is the best in very acute stroke patients? *Topics in Stroke Rehabilitation*, *26*(5), 359–365. doi:10.1080/10749357.2019.1607994 PMID:31044664

Bassetti, C. L. (2005). Sleep and stroke. *Seminars in Neurology*, *25*(1), 19–32. doi:10.1055-2005-867073 PMID:15798934

Bassetti, C. L., & Hermann, D. M. (2011). Sleep and stroke. In *Handbook of clinical neurology* (Vol. 99, pp. 1051–1072). Elsevier. doi:10.1016/B978-0-444-52007-4.00021-7

Basso, A., Forbes, M., & Boller, F. (2013). Rehabilitation of aphasia. In *Handbook of clinical neurology* (Vol. 110, pp. 325–334). Elsevier. doi:10.1016/B978-0-444-52901-5.00027-7

Bauer, C., & Girardi, M. (2002). Vestibular rehabilitation. *eMedicine Journal, 3*(6).

Bautmans, I., Jansen, B., Van Keymolen, B., & Mets, T. (2011). Reliability and clinical correlates of 3D-accelerometry based gait analysis outcomes according to age and fall-risk. *Gait & Posture*, *33*(3), 366–372. doi:10.1016/j.gaitpost.2010.12.003 PMID:21227697

Benou, I., & Raviv, T. R. (2019). DeepTract: A probabilistic Deep Learning Framework for White Matter FIber Tractography. *International Conference on medical image computing and computer-based intervention.* 10.1007/978-3-030-32248-9_70

Berman, J. I., Chung, S., Mukherjee, P., Hess, C. P., Han, E. T., & Henry, R. G. (2008). Probabilistic streamline q-ball tractography using the residual bootstrap. *NeuroImage*, *39*(1), 215–222. Advance online publication. doi:10.1016/j.neuroimage.2007.08.021 PMID:17911030

Berthier, M. L. (2005). Poststroke aphasia. *Drugs & Aging*, *22*(2), 163–182. doi:10.2165/00002512-200522020-00006 PMID:15733022

Bhatt, A., & Jani, V. (2011). The ABCD and ABCD2 Scores and the Risk of Stroke following a TIA: A Narrative Review. *ISRN Neurology*, *518621*(10), 21. doi:10.5402/2011/518621 PMID:22389822

Bigger, E., & Fraguada, L. (2016). Programmable Plaid: The Search for Seamless Integration in Fashion and Technology. *Proceedings of the 2016 ACM International Joint Conference on Pervasive and Ubiquitous Computing(UbiComp'16)*, 464–46. 10.1145/2968219.2971343

Bittar, R. S. M., & Barros, C. G. C. (2011). Vestibular rehabilitation with biofeedback in patients with central imbalance. *Revista Brasileira de Otorrinolaringologia (English Ed.)*, *77*(3), 356–361. doi:10.1590/S1808-86942011000300014 PMID:21739011

Bohannon, R. W. (1992). Lateral trunk flexion strength: Impairment, measurement reliability and implications following unilateral brain lesion. *International Journal of Rehabilitation Research. Internationale Zeitschrift fur Rehabilitationsforschung. Revue Internationale de Recherches de Readaptation*, *15*(3), 249–251. doi:10.1097/00004356-199209000-00010 PMID:1428392

Bohannon, R. W., Cassidy, D., & Walsh, S. (1995). Trunk muscle strength is impaired multidirectionally after stroke. *Clinical Rehabilitation*, *9*(1), 47–51. doi:10.1177/026921559500900107

Bohannon, R. W., & Smith, M. B. (1987). Interrater reliability of a modified Ashworth scale of muscle spasticity. *Physical Therapy*, *67*(2), 206–207. doi:10.1093/ptj/67.2.206 PMID:3809245

Bouziana, S. D., & Tziomalos, K. (2011). Malnutrition in patients with acute stroke. *Journal of Nutrition and Metabolism, 2011*, 1–7. doi:10.1155/2011/167898 PMID:22254136

Braun, V., & Clarke, V. (2006). Using thematic analysis in psychology. *Qualitative Research in Psychology, 3*(2), 77–101. doi:10.1191/1478088706qp063oa

Breining, B. L., & Sebastian, R. (2020). Neuromodulation in Post-stroke Aphasia Treatment. *Current Physical Medicine and Rehabilitation Reports, 8*(2), 44–56. doi:10.100740141-020-00257-5 PMID:33344066

Bryman, A. (2012). *Social Research Methods* (4th ed.). Oxford University Press.

Buğdaycı, D., & Caf, N. (2016). Poststroke cognitive impairments, apraxia and neglect. *Turkiye Klinikleri J PM&R-Special Topics, 9*(1), 17–23.

Burns, S. P., & Spanier, D. E. (2005). Break- technique hand held dynamometry: Relation between angular velocity and strength measurements. *Archives of Physical Medicine and Rehabilitation, 86*(7), 1420–1426. doi:10.1016/j.apmr.2004.12.041 PMID:16003675

Büyükavcı, R., Şahin, F., Sağ, S., Doğu, B., & Kuran, B. (2016). The impact of additional trunk balance exercises on balance, functional condition and ambulation in early stroke patients: Randomized controlled trial. *Turkish Journal of Physical Medicine and Rehabilitation, 62*(3), 248–256. doi:10.5606/tftrd.2016.84770

Caplan, L. R., & Kumar, S. (2016). Complications in stroke patients. In L. R. Caplan (Ed.), Caplan's Stroke: A Clinical Approach (5th ed., pp. 594-607). Cambridge University Press. doi:10.1017/CBO9781316095805.020

Carey, L. M., & Matyas, T. A. (2011). Frequency of discriminative sensory loss in the hand after stroke in a rehabilitation setting. *Journal of Rehabilitation Medicine, 43*(3), 257–263. doi:10.2340/16501977-0662 PMID:21305243

Carlsson, H., Gard, G., & Brogårdh, C. (2018). Upper-limb sensory impairments after stroke: Self-reported experiences of daily life and rehabilitation. *Journal of Rehabilitation Medicine, 50*(1), 45–51. doi:10.2340/16501977-2282 PMID:29068038

Carozzoa, S., Serraa, S., Pignoloa, L., Tonina, P., & Cerasa, A. (2020). The assessment of trunk recovery in stroke patients using 3D kinematic measures. *Medical Engineering & Physics, 78*, 98–105. doi:10.1016/j.medengphy.2020.01.013 PMID:32035812

Ceryan, K., & MB, Ş. (2007). Baş dönmesi olan hastada öykü, muayene ve tanı. *Kulak burun boğaz ve baş boyun cerrahisi, 2*, 36-63.

Çevikol, A., & Çakcı, A. (2015). İnme rehabilitasyonu [Stroke rehabilitation]. In H. Oğuz, H. Çakırbay, & B. Yanık (Eds.), *Tıbbi Rehabilitasyon* [Medical Rehabilitation] (pp. 419–448). Nobel Tıp Kitabevleri.

Chang, B. P., Rostanski, S., Willey, J., Kummer, B., Miller, E., & Elkind, M. (2018). Can I Send This Patient with Stroke Home? Strategies Managing Transient Ischemic Attack and Minor Stroke in the Emergency Department. *The Journal of Emergency Medicine*, *54*(5), 636–644. doi:10.1016/j.jemermed.2017.12.015 PMID:29321107

Chantraine, A., Baribeault, A., Uebelhart, D., & Gremion, G. (1999). Shoulder pain and dysfunction in hemiplegia: Effects of functional electrical stimulation. *Archives of Physical Medicine and Rehabilitation*, *80*(3), 328–331. doi:10.1016/S0003-9993(99)90146-6 PMID:10084443

Chechlacz, M., Rotshtein, P., Bickerton, W.-L., Hansen, P. C., Deb, S., & Humphreys, G. W. (2010). Separating neural correlates of allocentric and egocentric neglect: Distinct cortical sites and common white matter disconnections. *Cognitive Neuropsychology*, *27*(3), 277–303. doi:10.1080/02643294.2010.519699 PMID:21058077

Chen, Z., Venkat, P., Seyfried, D., Chopp, M., Yan, T., & Chen, J. (2017). Brain–heart interaction: Cardiac complications after stroke. *Circulation Research*, *121*(4), 451–468. doi:10.1161/CIRCRESAHA.117.311170 PMID:28775014

Chohan, S. A., Venkatesh, P. K., & How, C. H. (2019). Long-term complications of stroke and secondary prevention: An overview for primary care physicians. *Singapore Medical Journal*, *60*(12), 616–620. doi:10.11622medj.2019158 PMID:31889205

Cholewicki, J., & McGill, S. M. (1996). Mechanical stability of the in vivo lumbar spine: Implications for injury and chronic low back pain. *Clinical Biomechanics (Bristol, Avon)*, *11*(1), 1–15. doi:10.1016/0268-0033(95)00035-6 PMID:11415593

Colley, A., Pakanen, M., Koskinen, S., Mikkonen, K., & Häkkilä, J. (2016). Smart Handbag as a Wearable PublicDisplay–Exploring Concepts and User Perceptions. *Proceedings of the 7th Augmented Human International Conference (AH'16)*, 1.

Collin, C., & Wade, D. (1990). Assessing motor impairment after stroke: A pilot reliability study. *Journal ofNeurology, Neurosurgery, and Psychiatry*, *53*(7), 576–579. doi:10.1136/jnnp.53.7.576 PMID:2391521

Connell, L. A., Lincoln, N., & Radford, K. (2008). Somatosensory impairment after stroke: Frequency of different deficits and their recovery. *Clinical Rehabilitation*, *22*(8), 758–767. doi:10.1177/0269215508090674 PMID:18678576

Cournan, M. (2012). Bladder management in female stroke survivors: Translating research into practice. *Rehabilitation Nursing*, *37*(5), 220–230. doi:10.1002/rnj.054 PMID:22949275

Coutts, S. B. (2017). Diagnosis and Management of Transient Ischemic Attack. *Continuum*, *23*(1), 82-92.

Cram, J. R., Kasman, G. S., & Holtz, J. (1998). *Introduction to Surface Electromyography*. Aspen Publishers Inc.

Crowfoot, G. M., van der Riet, P. J., & Maguire, J. M. (2016). Living within stories: Exploring the experiences of people with transient ischemic attack. *Nursing & Health Sciences*, *18*(1), 52–57. doi:10.1111/nhs.12227 PMID:26275055

Cumming, T. B., Marshall, R. S., & Lazar, R. M. (2013). Stroke, cognitive deficits, and rehabilitation: Still an incomplete picture. *International Journal of Stroke*, *8*(1), 38–45. doi:10.1111/j.1747-4949.2012.00972.x PMID:23280268

Daniels, S. K., Brailey, K., Priestly, D. H., Herrington, L. R., Weisberg, L. A., & Foundas, A. L. (1998). Aspiration in patients with acute stroke. *Archives of Physical Medicine and Rehabilitation*, *79*(1), 14–19. doi:10.1016/S0003-9993(98)90200-3 PMID:9440410

Davies, P. M. (1990). Right in the middle. Springer-Verlag. doi:10.1007/978-3-642-61502-3

de Oliveira, F. F., Marin, S. M. C., & Bertolucci, P. H. F. (2013). Communicating with the non-dominant hemisphere: Implications for neurological rehabilitation. *Neural Regeneration Research*, *8*(13), 1236. doi:10.3969/j.issn.1673-5374.2013.13.009 PMID:25206418

Delpont, B., Blanc, C., Osseby, G., Hervieu-Bègue, M., Giroud, M., & Béjot, Y. (2018). Pain after stroke: A review. *Revue Neurologique*, *174*(10), 671–674. doi:10.1016/j.neurol.2017.11.011 PMID:30054011

Demeyere, N., & Gillebert, C. R. (2019). Ego-and allocentric visuospatial neglect: Dissociations, prevalence, and laterality in acute stroke. *Neuropsychology*, *33*(4), 490–498. doi:10.1037/neu0000527 PMID:30896236

Di Benedetto, A. C. (2017). Corporate social responsibility as emerging business model in fashion marketing. *J. Glob.Fash. Mark.*, *8*(4), 251–265. doi:10.1080/20932685.2017.1329023

Di Benedetto, P. (2011). Clean intermittent self-catheterization in neuro-urology. *European Journal of Physical and Rehabilitation Medicine*, *47*(4), 651–659. PMID:22222962

Dickstein, R., Heffes, Y., Laufer, Y., & Ben-Haim, Z. (1999). Activation of selected trunk muscles during symmetric functional activities in poststrokehemiparetic and hemiplegic patients. *Journal of Neurology, Neurosurgery, and Psychiatry*, *66*(2), 218–221. doi:10.1136/jnnp.66.2.218 PMID:10071103

Dickstein, R., Shefi, S., Marcovitz, E., & Villa, Y. (2004). Anticipatory postural adjustment in selected trunk muscles in post stroke hemiparetic patients. *Archives of Physical Medicine and Rehabilitation*, *85*(2), 261–267. doi:10.1016/j.apmr.2003.05.011 PMID:14966711

Dirnagl, U., Becker, K., & Meisel, A. (2009). Preconditioning and tolerance against cerebral ischaemia: From experimental strategies to clinical use. *Lancet Neurology*, *8*(4), 398–412. doi:10.1016/S1474-4422(09)70054-7 PMID:19296922

Dix, M. (1974). Treatment of vertigo. *Physiotherapy*, *60*(12), 380–384. PMID:4467199

Donnan, G. A., & Norrving, B. (2008). Lacunes and lacunar syndromes. *Handbook of Clinical Neurology*, *93*, 559–575. doi:10.1016/S0072-9752(08)93027-X PMID:18804668

Donovan, N. J., Kendall, D. L., Heaton, S. C., Kwon, S., Velozo, C. A., & Duncan, P. W. (2008). Conceptualizing functional cognition in stroke. *Neurorehabilitation and Neural Repair*, *22*(2), 122–135. doi:10.1177/1545968307306239 PMID:17761809

Dorsher, P. T., & McIntosh, P. M. (2012). Neurogenic bladder. *Advances in Urology*, *816274*, 1–16. Advance online publication. doi:10.1155/2012/816274 PMID:22400020

Durlanık, G. (2016). Vascular anatomy and correlation with the clinical features by stroke. *Turkiye Klinikleri J PM&R-Special Topics*, *9*(1), 1–7.

Easton, J. D., Saver, J. L., Albers, G. W., Alberts, M. J., Chaturvedi, S., Feldmann, E., ... Kidwell, C. S. (2009). Definition and evaluation of transient ischemic attack: a scientific statement for healthcare professionals from the American Heart Association/American Stroke Association Stroke Council; Council on Cardiovascular Surgery and Anesthesia; Council on Cardiovascular Radiology and Intervention; Council on Cardiovascular Nursing; and the Interdisciplinary Council on Peripheral Vascular Disease: the American Academy of Neurology affirms the value of this statement as an educational tool for neurologists. *Stroke*, *40*(6), 2276–2293. doi:10.1161/STROKEAHA.108.192218 PMID:19423857

Edlow, J. A. (2018). Managing Patients With Transient Ischemic Attack. *Annals of Emergency Medicine*, *71*(3), 409–415. doi:10.1016/j.annemergmed.2017.06.026 PMID:28754355

Edlow, J. A., Kim, S., Pelletier, A. J., & Camargo, C. A. Jr. (2006). National study on emergency department visits for transient ischemic attack, 1992-2001. *Academic Emergency Medicine*, *13*(6), 666–672. PMID:16609106

Ferber, S., & Karnath, H.-O. (2001). How to assess spatial neglect-line bisection or cancellation tasks? *Journal of Clinical and Experimental Neuropsychology*, *23*(5), 599–607. doi:10.1076/jcen.23.5.599.1243 PMID:11778637

Ferro, J. M. (2001). Hyperacute cognitive stroke syndromes. *Journal of Neurology*, *248*(10), 841–849. doi:10.1007004150170067 PMID:11697519

Ferro, J. M. (2003). Cardioembolic stroke: An update. *Lancet Neurology*, *2*(3), 177–188. doi:10.1016/S1474-4422(03)00324-7 PMID:12849239

Fletcher, B. J., Dunbar, S., Coleman, J., Jann, B., & Fletcher, G. F. (1993). Cardiac precautions for non-acute inpatient settings. *American Journal of Physical Medicine & Rehabilitation*, *72*(3), 140–143. doi:10.1097/00002060-199306000-00006 PMID:8512675

Francisco, G. E., & McGuire, J. R. (2012). Poststroke spasticity management. *Stroke*, *43*(11), 3132–3136. doi:10.1161/STROKEAHA.111.639831 PMID:22984012

Frizzell, J. P. (2005). Acute stroke: Pathophysiology, diagnosis, and treatment. *AACN Advanced Critical Care*, *16*(4), 421–440. doi:10.1097/00044067-200510000-00002 PMID:16269890

Fujiwara, T., Liu, M., Tsuji, T., Sonoda, S., Mizuno, K., Akaboshi, K., Hase, K., Masakado, Y., & Chino, N. (2004). Development of a new measure to assess trunk impairment after stroke (trunk impairment scale): Its psychometric properties. *American Journal of Physical Medicine & Rehabilitation*, *83*(9), 681–688. doi:10.1097/01.PHM.0000137308.10562.20 PMID:15314532

Ganzer, C. A., Barnes, A., Uphold, C., & Jacobs, A. R. (2016). Transient ischemic attack and cognitive impairment: A review. *The Journal of Neuroscience Nursing*, *48*(6), 322–327. doi:10.1097/JNN.0000000000000236 PMID:27824800

Geurts, A., Visschers, B., Van Limbeek, J., & Ribbers, G. (2000). Systematic review of aetiology and treatment of post-stroke hand oedema and shoulder-hand syndrome. *Scandinavian Journal of Rehabilitation Medicine*, *32*(1), 4–10. doi:10.1080/003655000750045668 PMID:10782934

Glasscock, M. E., Cueva, R. A., & Thedinger, B. A. (1990). *Handbook of vertigo*. Raven Press.

GlobalF. A. T. (2020). Available online: https://globalfashionagenda.com/commitment/#toolboxes

GlobalO. T. S. (2019). Available online: https://www.global-standard.org/the-standard/general-description.html

Gordon, N. F., Gulanick, M., Costa, F., Fletcher, G., Franklin, B. A., Roth, E. J., & Shephard, T. (2004). Physical activity and exercise recommendations for stroke survivors: An American Heart Association scientific statement from the Council on Clinical Cardiology, Subcommittee on Exercise, Cardiac Rehabilitation, and Prevention; the Council on Cardiovascular Nursing; the Council on Nutrition, Physical Activity, and Metabolism; and the Stroke Council. *Circulation*, *109*(16), 2031–2041. doi:10.1161/01.CIR.0000126280.65777.A4 PMID:15117863

Granacher, U., Lacroix, A., Muehlbauer, T., Roettger, K., & Gollhofer, A. (2013). Effects of core instability strength training on trunk muscle strength, spinal mobility, dynamic balance and functional mobility in older adults. *Gerontology*, *59*(2), 105–113. doi:10.1159/000343152 PMID:23108436

Hackam, D. G., & Spence, J. D. (2007). Combining multiple approaches for the secondary prevention of vascular events after stroke: A quantitative modeling study. *Stroke*, *38*(6), 1881–1885. doi:10.1161/STROKEAHA.106.475525 PMID:17431209

Halmagyi, G. M., & Cremer, P. (2000). Assessment and treatment of dizziness. BMJ Publishing Group Ltd.

Hankey, G. J., Johnston, S. C., Easton, J. D., Hacke, W., Mas, J. L., Brennan, D., Mak, K. H., Bhatt, D. L., Fox, K. A. A., & Topol, E. J. (2011). Effect of clopidogrel plus ASA vs. ASA early after TIA and ischaemic stroke: A substudy of the CHARISMA trial. *International Journal of Stroke*, *6*(1), 3–9. doi:10.1111/j.1747-4949.2010.00535.x PMID:21205234

Harden, R. N. (2005). Pharmacotherapy of complex regional pain syndrome. *American Journal of Physical Medicine & Rehabilitation*, *84*(3), S17–S28. doi:10.1097/01.PHM.0000154906.87708.C9 PMID:15722780

Harden, R. N., Oaklander, A. L., Burton, A. W., Perez, R. S., Richardson, K., Swan, M., Barthel, J., Costa, B., Graciosa, J. R., & Bruehl, S. (2013). Complex regional pain syndrome: Practical diagnostic and treatment guidelines. *Pain Medicine*, *14*(2), 180–229. doi:10.1111/pme.12033 PMID:23331950

Harrison, R. A., & Field, T. S. (2015). Post stroke pain: Identification, assessment, and therapy. *Cerebrovascular Diseases (Basel, Switzerland)*, *39*(3-4), 190–201. doi:10.1159/000375397 PMID:25766121

Hart, R. G., Pearce, L. A., & Aguilar, M. I. (2007). Meta-analysis: Antithrombotic therapy to prevent stroke in patients who have nonvalvular atrial fibrillation. *Annals of Internal Medicine*, *146*(12), 857–867. doi:10.7326/0003-4819-146-12-200706190-00007 PMID:17577005

Harvey, R., Roth, E., & Yu, D. (2007). Rehabilitation in stroke syndromes. In R. L. Braddom (Ed.), *Physical Medicine & Rehabilitation* (pp. 1175–1112). Saunders Elsevier.

Hecht, J. S. (1992). Subscapular nerve block in the painful hemiplegic shoulder. *Archives of Physical Medicine and Rehabilitation*, *73*(11), 1036–1039. PMID:1444768

Heijenbrok-Kal, M. H., Buskens, E., Nederkoorn, P. J., van der Graaf, Y., & Hunink, M. G. (2006). Optimal peak systolic velocity threshold at duplex us for determining the need for carotid endarterectomy: A decision analytic approach. *Radiology*, *238*(2), 480–488. doi:10.1148/radiol.2381041078 PMID:16371575

Hischier, R., Coroama, V. C., Schien, D., & Achachlouei, M. A. (2015). Grey energy and environmental impacts of ICThardware. In ICT Innovations for Sustainability. Springer.

Hunt, C. P., Wickham, M., Pittson, R., & Lewison, J. (2018). *Reusable, Unzippable, Sustainable Electronics (ReUse) Interconnect System for the Circular Economy*. Report for NPL, MAT 75. Available online: https://eprintspublications.npl.co.uk/6645/

Hurford, R., Charidimou, A., Fox, Z., Cipolotti, L., & Werring, D. J. (2013). Domain-specific trends in cognitive impairment after acute ischaemic stroke. *Journal of Neurology*, *260*(1), 237–241. doi:10.100700415-012-6625-0 PMID:22865200

Hwang, W. J., Cho, M. K., & Chung, Y. (2015). Relationship between anticipatory postural adjustment of the trunk, dual tasks and physical performance with chronic stroke survivors: A pilot test. *Physical Therapy Rehabilitation Science*, *4*(1), 44–48. doi:10.14474/ptrs.2015.4.1.44

Indredavik, B., Rohweder, G., Naalsund, E., & Lydersen, S. (2008). Medical complications in a comprehensive stroke unit and an early supported discharge service. *Stroke*, *39*(2), 414–420. doi:10.1161/STROKEAHA.107.489294 PMID:18096834

Iosa, M., Fusco, A., Morone, G., Pratesi, L., Coiro, P., Venturiero, V., De Angelis, D., Bragoni, M., & Paolucci, S. (2012). Assessment of upper-body dynamic stability during walking in patients with subacute stroke. *Journal of Rehabilitation Research and Development*, *49*(3), 439–450. doi:10.1682/JRRD.2011.03.0057 PMID:22773202

Iurato, L., Randisi, G., Torrisi, J., Grimaldi, R., Naso, G., Verniccio, S., Occhipinti, C., & Vecchio, M. M. (2015). A temible complication of ischemic stroke: Pulmonary embolism. *Clinical Management Issues*, *9*(4), 101–108. doi:10.7175/cmi.v9i4.1196

Jacobson, G. P., Newman, C. W., & Kartush, J. M. (1993). *Handbook of balance function testing*. Mosby Elsevier Health Science.

Johnston, K. C., Li, J. Y., Lyden, P. D., Hanson, S. K., Feasby, T. E., Adams, R. J., Faught, R. E. Jr, & Haley, E. C. Jr. (1998). Medical and neurological complications of ischemic stroke: Experience from the RANTTAS trial. *Stroke*, *29*(2), 447–453. doi:10.1161/01.STR.29.2.447 PMID:9472888

Johnston, S. C., Fayad, P. B., Gorelick, P. B., Hanley, D. F., Shwayder, P., van Husen, D., & Weiskopf, T. (2003). Prevalence and knowledge of transient ischemic attack among US adults. *Neurology*, *60*(9), 1429–1434. doi:10.1212/01.WNL.0000063309.41867.0F PMID:12743226

Johnston, S. C., Gress, D. R., Browner, W. S., & Sidney, S. (2000). Short-term prognosis after emergency department diagnosis of TIA. *Journal of the American Medical Association*, *284*(22), 2901–2906. doi:10.1001/jama.284.22.2901 PMID:11147987

Johnston, S. C., Rothwell, P. M., Nguyen-Huynh, M. N., Giles, M. F., Elkins, J. S., Bernstein, A. L., & Sidney, S. (2007). Validation and refinement of scores to predict very early stroke risk after transient ischaemic attack. *Lancet*, *369*(9558), 283–292. doi:10.1016/S0140-6736(07)60150-0 PMID:17258668

Jönsson, A.-C., Lindgren, I., Hallström, B., Norrving, B., & Lindgren, A. (2006). Prevalence and intensity of pain after stroke: A population based study focusing on patients' perspectives. *Journal of Neurology, Neurosurgery, and Psychiatry*, *77*(5), 590–595. doi:10.1136/jnnp.2005.079145 PMID:16354737

Josephson, S. A., Bryant, S. O., Mak, H. K., Johnston, S. C., Dillon, W. P., & Smith, W. S. (2004). Evaluation of carotid stenosis using CT angiography in the initial evaluation of stroke and TIA. *Neurology*, *63*(3), 457–460. doi:10.1212/01.WNL.0000135154.53953.2C PMID:15304575

Kamal, A. K., Naqvi, I., Husain, M. R., & Khealani, B. A. (2011). Cilostazol versus aspirin for secondary prevention of vascular events after stroke of arterial origin. *Cochrane Database of Systematic Reviews*, *19*(1). PMID:21249700

Karataş, G. (2011). İnme [Stroke]. In M. Beyazova & Y. G. Kutsal (Eds.), *Fiziksel Tıp ve Rehabilitasyon* [Physical Medicine and Rehabilitation] (pp. 2761–2783). Güneş Kitabevi.

Karthikbabu, S., & Chakrapani, M. (2017). Hand-Held Dynamometer is a Reliable Tool to Measure Trunk Muscle Strength in Chronic Stroke. *Journal of Clinical and Diagnostic Research: JCDR*, *11*(9), YC09–YC12. doi:10.7860/JCDR/2017/28105.10672 PMID:29207821

Karthikbabu, S., Chakrapani, M., Ganeshan, S., Rakshith, K. C., Nafeez, S., & Prem, V. (2012). A review on assessment and treatment of the trunk in stroke: A need or luxury. *Neural Regeneration Research*, *7*(25), 1974–1977. PMID:25624827

Keim, R. J., Cook, M., & Martini, D. (1992). How i do it: Otology and neurotology: A specific issue and its solution: Balance rehabilitation therapy. *The Laryngoscope*, *102*(11), 1302–1307. doi:10.1288/00005537-199211000-00019 PMID:1405995

Kelly, J., Rudd, A., Lewis, R., & Hunt, B. (2001). Venous thromboembolism after acute stroke. *Stroke*, *32*(1), 262–267. doi:10.1161/01.STR.32.1.262 PMID:11136946

Kendall, R. (2010). Musculoskeletal problems in stroke survivors. *Topics in Stroke Rehabilitation*, *17*(3), 173–178. doi:10.1310/tsr1703-173 PMID:20797960

Kennedy, J., Hill, M. D., Ryckborst, K. J., Eliasziw, M., Demchuk, A. M., & Buchan, A. M. (2007). Fast assessment of stroke and transient ischaemic attack to prevent early recurrence (FASTER): A randomised controlled pilot trial. *Lancet Neurology*, *6*(11), 961–969. doi:10.1016/S1474-4422(07)70250-8 PMID:17931979

Kernan, W. N., Ovbiagele, B., Black, H. R., Bravata, D. M., Chimowitz, M. I., Ezekowitz, M. D., Fang, M. C., Fisher, M., Furie, K. L., Heck, D. V., Johnston, S. C. C., Kasner, S. E., Kittner, S. J., Mitchell, P. H., Rich, M. W., Richardson, D. J., Schwamm, L. H., & Wilson, J. A. (2014). Guidelines for the prevention of stroke in patients with stroke and transient ischemic attack: A guideline for healthcare professionals from the American Heart Association/American Stroke Association. *Stroke*, *45*(7), 2160–2236. doi:10.1161/STR.0000000000000024 PMID:24788967

Kernan, W. N., Viscoli, C. M., Furie, K. L., Young, L. H., Inzucchi, S. E., Gorman, M., Guarino, P. D., Lovejoy, A. M., Peduzzi, P. N., Conwit, R., Brass, L. M., Schwartz, G. G., Adams, H. P. Jr, Berger, L., Carolei, A., Clark, W., Coull, B., Ford, G. A., Kleindorfer, D., ... Winder, T. R. (2016). Pioglitazone after Ischemic Stroke or Transient Ischemic Attack. *The New England Journal of Medicine*, *374*(14), 1321–1331. doi:10.1056/NEJMoa1506930 PMID:26886418

Khan, M. T., Ikram, A., Saeed, O., Afridi, T., Sila, C. A., Smith, M. S., ... Shuaib, A. (2017). Deep vein thrombosis in acute stroke-a systemic review of the literature. *Cureus*, *9*(12). Advance online publication. doi:10.7759/cureus.1982 PMID:29503776

Khare, S. (2016). Risk factors of transient ischemic attack: An overview. *Journal of Midlife Health*, *7*(1), 2–7. doi:10.4103/0976 7800.179166 PMID:27134474

Kim, J. S., Kwon, S. U., & Uchiyama, S. (2015). Cilostazol research in Asia: Can it be applied to European and American patients? *International Journal of Stroke*, *1*(1_suppl), 1–9. doi:10.1111/ijs.12460 PMID:25736819

Kiyohara, T., Kamouchi, M., Kumai, Y., Ninomiya, T., Hata, J., Yoshimura, S., ... Kitazono, T. (2014). ABCD3 and ABCD3-I scores are superior to ABCD2 score in the prediction of short- and long-term risks of stroke after transient ischemic attack. *Stroke*, *45*(2), 418–425. doi:10.1161/STROKEAHA.113.003077 PMID:24335223

Köhler, A., & Erdmann, L. (2004). Expected environmental impacts of pervasive computing. *Hum. Ecol. Risk Assess*, *10*, 831–852.

Koog, Y. H., Jin, S. S., Yoon, K., & Min, B.-I. (2010). Interventions for hemiplegic shoulder pain: Systematic review of randomised controlled trials. *Disability and Rehabilitation, 32*(4), 282–291. doi:10.3109/09638280903127685 PMID:20055567

Koyuncu, E., & Özgirgin, N. (2016). Neurogenic language and speech disorders and rehabilitation in stroke. *Turkiye Klinikleri J PM&R-Special Topics, 9*(1), 8–16.

Krishnan, K., Beridze, M., Christensen, H., Dineen, R., Duley, L., Heptinstall, S., ... Bath, P. (2015). Safety and efficacy of intensive vs. guideline antiplatelet therapy in high-risk patients with recent ischemic stroke or transient ischemic attack: Rationale and design of the Triple Antiplatelets for Reducing Dependency after Ischaemic Stroke (TARDIS) trial (ISRCTN47823388). *International Journal of Stroke, 10*(7), 1159–1165. doi:10.1111/ijs.12538 PMID:26079743

Kumar, S., Selim, M. H., & Caplan, L. R. (2010). Medical complications after stroke. *Lancet Neurology, 9*(1), 105–118. doi:10.1016/S1474-4422(09)70266-2 PMID:20083041

Lauren, J. (2011). An introduction to diffusion tensor image analysis. *Neurosurgery Clinics of North America, 22*(2), 185–196. doi:10.1016/j.nec.2010.12.004 PMID:21435570

Lazar, R. M., & Boehme, A. K. (2017). Aphasia as a predictor of stroke outcome. *Current Neurology and Neuroscience Reports, 17*(11), 83. doi:10.100711910-017-0797-z PMID:28929424

Lee, M. J., Lee, J. H., Koo, H. M., & Lee, S. M. (2017). Effectiveness of bilateral arm training for improving extremity function and activities of daily living performance in hemiplegic patients. *Journal of Stroke and Cerebrovascular Diseases, 26*(5), 1020–1025. doi:10.1016/j.jstrokecerebrovasdis.2016.12.008 PMID:28162905

Lee, N. G., You, J. S. H., Chung, H. Y., Jeon, H. S., Choi, B. S., Lee, D. R., & Yoon, H. S. (2018). Best Core Stabilization for Anticipatory Postural Adjustment and Falls in Hemiparetic Stroke. *Archives of Physical Medicine and Rehabilitation, 99*(11), 2168–2174. doi:10.1016/j.apmr.2018.01.027 PMID:29476713

Levy, D. E. (1988). How transient are transient ischemic attacks? *Neurology, 38*(5), 674–677. doi:10.1212/WNL.38.5.674 PMID:3362360

Lewis, R. P., & Kriukelyte, I. (2019). Complex neuropathic pain states. *Anaesthesia and Intensive Care Medicine, 20*(10), 579–582. doi:10.1016/j.mpaic.2019.07.013

Liao, C. F., Liaw, L. J., Wang, R. Y., Su, F. C., & Hsu, A. T. (2015). Electromyography of symmetrical trunk movements and trunk position sense in chronic stroke patients. *Journal of Physical Therapy Science, 27*(9), 2675–2681. doi:10.1589/jpts.27.2675 PMID:26504267

Lin, Y., Wang, A., Li, J., Lin, J., Wang, D., Meng, X., Ou, L., Chen, W., Zhao, X., Liu, L., Wang, Y., & Wang, Y. (2017). Impact of Glycemic Control on Efficacy of Clopidogrel in Transient Ischemic Attack or Minor Stroke Patients With CYP2C19 Genetic Variants. *Stroke, 48*(4), 998–1004. doi:10.1161/STROKEAHA.116.016463 PMID:28289237

Li, S., Francisco, G. E., & Zhou, P. (2018). Post-stroke hemiplegic gait: New perspective and insights. *Frontiers in Physiology, 9*, 1021. doi:10.3389/fphys.2018.01021 PMID:30127749

Liu, J., Feng, W., Zhou, J., Huang, F., Long, L., Wang, Y., Liu, P., Huang, X., Yang, M., Wang, K., & Sun, Z. (2020). Effects of sling exercise therapy on balance, mobility, activities of daily living, quality of life and shoulder pain in stroke patients: A randomized controlled trial. *European Journal of Integrative Medicine*, *35*(5), 1010772. doi:10.1016/j.eujim.2020.101077

Lovett, J. K., Dennis, M. S., Sandercock, P. A., Bamford, J., Warlow, C. P., & Rothwell, P. M. (2003). Very early risk of stroke after a first transient ischemic attack. *Stroke*, *34*(8), 10. doi:10.1161/01.STR.0000080935.01264.91 PMID:12855835

Maffiluetti, N. A., Bizzini, M., Desbrosses, K., Babault, N., & Munzinger, U. (2007). Relialibilty of knee extension and flexion measurements using the Con-Trex isokinetic dynamometer. *Clinical Physiology and Functional Imaging*, *27*(6), 346–353. doi:10.1111/j.1475-097X.2007.00758.x PMID:17944656

Malouin, F., Pichard, L., Bonneau, C., Durand, A., & Corriveau, D. (1994). Evaluating motor recovery early after stroke: Comparison of the Fugl-Meyer Assessment and the Motor Assessment Scale. *Archives of Physical Medicine and Rehabilitation*, *75*(11), 1206–1212. doi:10.1016/0003-9993(94)90006-X PMID:7979930

Mari, F., Matei, M., Ceravolo, M. G., Pisani, A., Montesi, A., & Provinciali, L. (1997). Predictive value of clinical indices in detecting aspiration in patients with neurological disorders. *Journal of Neurology, Neurosurgery, and Psychiatry*, *63*(4), 456–460. doi:10.1136/jnnp.63.4.456 PMID:9343123

Markus, H. (2008). Stroke: Causes and clinical features. *Medicine*, *36*(11), 586–591. doi:10.1016/j.mpmed.2008.08.009

McKenna, L. B. (2001). Hemiplegic shoulder pain: Defining the problem and its management. *Disability and Rehabilitation*, *23*(16), 698–705. doi:10.1080/09638280110062149 PMID:11732559

Mehdi, Z., Birns, J., & Bhalla, A. (2013). Post-stroke urinary incontinence. *International Journal of Clinical Practice*, *67*(11), 1128–1137. doi:10.1111/ijcp.12183 PMID:23834208

Mehndiratta, P., Smith, S. C., & Worrall, B. B. (2015). Etiologic stroke subtypes: updated definition and efficient workup strategies. *Current Treatment Options in Cardiovascular Medicine, 17*(1), 1-11. doi:10.1007/s11936-014-0357-7

Melo, T. P., & Bogousslavsky, J. (2001). Hemiparesis and other types of motor weakness. In J. Bogousslavsky & L. R. Caplan (Eds.), *Stroke Syndromes* (pp. 22–33). Cambridge University Press. doi:10.1017/CBO9780511586521.004

Merwick, A., Albers, G. W., Amarenco, P., Arsava, E. M., Ay, H., Calvet, D., Coutts, S. B., Cucchiara, B. L., Demchuk, A. M., Furie, K. L., Giles, M. F., Labreuche, J., Lavallée, P. C., Mas, J.-L., Olivot, J. M., Purroy, F., Rothwell, P. M., Saver, J. L., Sheehan, Ó. C., ... Kelly, P. J. (2010). Addition of brain and carotid imaging to the $ABCD^2$ score to identify patients at early risk of stroke after transient ischaemic attack: A multicentre observational study. *Lancet Neurology*, *9*(11), 1060–1069. doi:10.1016/S1474-4422(10)70240-4 PMID:20934388

Nadarajan, V., Perry, R. J., Johnson, J., & Werring, D. J. (2014). Transient ischaemic attacks: Mimics and chameleons. *Practical Neurology*, *14*(1), 23–31. doi:10.1136/practneurol-2013-000782 PMID:24453269

National Institute for Health Care Excellence. (2018). *Venous Thromboembolism in over 16s: reducing the risk of hospital-acquired deep vein thrombosis or pulmonary embolism.* NICE Guideline [NG89].

Nazliel, B. (2018). *Transient Ischemic Attack. Gokce M, editör. Stroke – I.* Turkish Clinics.

Neher, Gotz, Norajitra, Weber, & Maier-Hein. (2015). A machine learning based approach to Fiber tractography using Classifier Voting. Medical Image Computing and Computer-Assisted Intervention (MICCAI), 45–52.

Nelson, L. A. (2007). The role of biofeedback in stroke rehabilitation: Past and future directions. *Topics in Stroke Rehabilitation*, *14*(4), 59–66. doi:10.1310/tsr1404-59 PMID:17698458

Onursal, F. Ö. (2015). *Inmede Gövde Eğitimi. Fizyoterapistler ve öğrenciler için e-kitap Bahar Dönemi.* https://www.researchgate.net/publication/329717723

Osaka, H., Shinkoda, K., Watanabe, S., Fujita, D., Kobara, K., Yoshimura, Y., Ito, T., & Suehiro, T. (2017). Association between trunk acceleration during walking and clinically assessed balance in patients with stroke. *NeuroRehabilitation*, *41*(4), 783–790. doi:10.3233/NRE-172171 PMID:29254113

Osawa, A., Maeshima, S., & Tanahashi, N. (2013). Water-swallowing test: Screening for aspiration in stroke patients. *Cerebrovascular Diseases (Basel, Switzerland)*, *35*(3), 276–281. doi:10.1159/000348683 PMID:23548854

Ostaszkiewicz, J., Johnston, L., & Roe, B. (2004). Timed voiding for the management of urinary incontinence in adults. *Cochrane Database of Systematic Reviews*, (1), 002802. doi:10.1002/14651858.CD002802.pub2 PMID:14973993

Paci, M., Nannetti, L., & Rinaldi, L. A. (2005). Glenohumeral subluxation in hemiplegia: An overview. *Journal of Rehabilitation Research and Development*, *42*(4), 557. Advance online publication. doi:10.1682/JRRD.2004.08.0112 PMID:16320150

Palsetia, D., Rao, G. P., Tiwari, S. C., Lodha, P., & De Sousa, A. (2018). The clock drawing test versus mini-mental status examination as a screening tool for dementia: A clinical comparison. *Indian Journal of Psychological Medicine*, *40*(1), 1–10. doi:10.4103/IJPSYM.IJPSYM_244_17 PMID:29403122

Paolucci, S., Antonucci, G., Grasso, M. G., Morelli, D., Troisi, E., Coiro, P., De Angelis, D., Rizzi, F., & Bragoni, M. (2001). Post-stroke depression, antidepressant treatment and rehabilitation results. *Cerebrovascular Diseases (Basel, Switzerland)*, *12*(3), 264–271. doi:10.1159/000047714 PMID:11641594

Pare, J. R., & Kahn, J. H. (2012). Basic neuroanatomy and stroke syndromes. *Emergency Medicine Clinics*, *30*(3), 601–615. doi:10.1016/j.emc.2012.05.004 PMID:22974640

Park, J. E. (2017). Apraxia: Review and update. *Journal of Clinical Neurology (Seoul, Korea)*, *13*(4), 317–324. doi:10.3988/jcn.2017.13.4.317 PMID:29057628

Park, S., & Park, J. Y. (2016). Grip strength in post-stroke hemiplegia. *Journal of Physical Therapy Science*, *28*(2), 677–679. doi:10.1589/jpts.28.677 PMID:27065562

Pedersen, P. M., Vinter, K., & Olsen, T. S. (2004). Aphasia after stroke: Type, severity and prognosis. *Cerebrovascular Diseases (Basel, Switzerland)*, *17*(1), 35–43. doi:10.1159/000073896 PMID:14530636

Phan, T., Huston, J. III, Bernstein, M. A., Riederer, S. J., & Brown, R. D. Jr. (2001). Contrast-enhanced magnetic resonance angiography of the cervical vessels: Experience with 422 patients. *Stroke*, *32*(10), 2282–2286. doi:10.1161/hs1001.096046 PMID:11588314

Pigeon, P., Yahia, L. H., Mitnitski, A. B., & Feldman, A. G. (2000). Superposition of independent units of coordination during pointing movements involving the trunk with and without visual feedback. *Experimental Brain Research*, *131*(3), 336–349. doi:10.1007002219900267 PMID:10789948

Poulin, P., Cote, M.-A., Houde, J.-C., Petit, L., Neher, P. F., Maier-Hein, K. H., Larochelle, H., & Descoteaux, M. (2017). Learn to track: Deep learning for tractography. Medical Image Computing and Computer-Assisted Intervention (MICCAI), 540–547. doi:10.1101/146688

Pretorius, S., & Van Zyl, L. (2008). Nutritional therapy in stroke management: prevention and rehabilitation: nutrition. *CME: Your SA Journal of CPD, 26*(3), 164-166. https://hdl.handle.net/10520/EJC63446

Ramsey, D. J., Smithard, D. G., & Kalra, L. (2003). Early assessments of dysphagia and aspiration risk in acute stroke patients. *Stroke*, *34*(5), 1252–1257. doi:10.1161/01.STR.0000066309.06490.B8 PMID:12677020

Ramsey, D., Smithard, D., & Kalra, L. (2005). Silent aspiration: What do we know? *Dysphagia*, *20*(3), 218–225. doi:10.100700455-005-0018-9 PMID:16362510

Robinson, R. G., & Jorge, R. E. (2016). Post-stroke depression: A review. *The American Journal of Psychiatry*, *173*(3), 221–231. doi:10.1176/appi.ajp.2015.15030363 PMID:26684921

Robinson, R. G., & Spalletta, G. (2010). Poststroke depression: A review. *Canadian Journal of Psychiatry*, *55*(6), 341–349. doi:10.1177/070674371005500602 PMID:20540828

Román, G. C. (2003). Vascular dementia: distinguishing characteristics, treatment, and prevention. *Journal of the American Geriatrics Society*, *51*(5s2), S296-S304. doi:10.1046/j.1532-5415.5155.x

Rothwell, P. M., Giles, M. F., Chandratheva, A., Marquardt, L., Geraghty, O., Redgrave, J. N., Lovelock, C. E., Binney, L. E., Bull, L. M., Cuthbertson, F. C., Welch, S. J. V., Bosch, S., Carasco-Alexander, F., Silver, L. E., Gutnikov, S. A., & Mehta, Z. (2007). Effect of urgent treatment of transient ischaemic attack and minor stroke on early recurrent stroke (EXPRESS study): A prospective population-based sequential comparison. *Lancet*, *370*(9596), 1432–1442. doi:10.1016/S0140-6736(07)61448-2 PMID:17928046

Rothwell, P. M., Giles, M. F., Flossmann, E., Lovelock, C. E., Redgrave, J. N., Warlow, C. P., & Mehta, Z. (2005). A simple score (ABCD) to identify individuals at high early risk of stroke after transient ischaemic attack. *Lancet*, *366*(9479), 29–36. doi:10.1016/S0140-6736(05)66702-5 PMID:15993230

Sahathevan, R., Brodtmann, A., & Donnan, G. A. (2012). Dementia, stroke, and vascular risk factors; a review. *International Journal of Stroke*, *7*(1), 61–73. doi:10.1111/j.1747-4949.2011.00731.x PMID:22188853

Salter, K., Teasell, R., Foley, N., & Allen, L. (2013). *Evidence-Based Review of Stroke Rehabilitation: Cognitive Disorders and Apraxia*. Academic Press.

Sandin, K. J., & Smith, B. S. (1990). The measure of balance in sitting in stroke rehabilitation prognosis. *Stroke*, *21*(1), 82–86. doi:10.1161/01.STR.21.1.82 PMID:2300995

Scheitz, J. F., Nolte, C. H., Doehner, W., Hachinski, V., & Endres, M. (2018). Stroke–heart syndrome: Clinical presentation and underlying mechanisms. *Lancet Neurology*, *17*(12), 1109–1120. doi:10.1016/S1474-4422(18)30336-3 PMID:30509695

Schoen, J. C., Boysen, M. M., Warren, C. R., Chakravarthy, B., & Lotfipour, S. (2011). Vertebrobasilar artery occlusion. *The Western Journal of Emergency Medicine*, *12*(2), 233. PMID:21691534

Sharma, V., & Kaur, J. (2017). Effect of core strengthening with pelvic proprioceptive neuromuscular facilitation on trunk, balance, gait, and function in chronic stroke. *Journal of Exercise Rehabilitation*, *13*(2), 200–205. doi:10.12965/jer.1734892.446 PMID:28503533

Sharpe, J. A., & Barber, H. O. (1993). *The vestibulo-ocular reflex and vertigo*. Lippincott Williams & Wilkins.

Shen, S.-X., & Liu, Y. (2018). A retrospective study of neuromuscular electrical stimulation for treating women with post-stroke incontinence. *Medicine*, *97*(26), e11264. Advance online publication. doi:10.1097/MD.0000000000011264 PMID:29952999

Shepard, N. T., Smith-Wheelock, M., Telian, S. A., & Raj, A. (1993). Vestibular and balance rehabilitation therapy. *The Annals of Otology, Rhinology, and Laryngology*, *102*(3), 198–205. doi:10.1177/000348949310200306 PMID:8457121

Shi, Y., Yang, D., Zeng, Y., & Wu, W. (2017). Risk factors for post-stroke depression: A meta-analysis. *Frontiers in Aging Neuroscience*, *9*, 218. doi:10.3389/fnagi.2017.00218 PMID:28744213

Shumway-Cook, A., & Woollacott, M. H. (2016). Motor Control: Translating Research Into *Clinics and Practice*. Wolters Kluwer.

Siket, M. S., & Edlow, J. (2013). Transient ischemic attack: An evidence-based update. *Emergency Medicine Practice*, *15*(1), 1–26. PMID:23257070

Smithard, D., O'neill, P., Park, C., Morris, J., Wyatt, R., England, R., & Martin, D. (1996). Complications and outcome after acute stroke: Does dysphagia matter? *Stroke*, *27*(7), 1200–1204. doi:10.1161/01.STR.27.7.1200 PMID:8685928

Smith, H. A., Lee, S. H., O'Neill, P. A., & Connolly, M. J. (2000). The combination of bedside swallowing assessment and oxygen saturation monitoring of swallowing in acute stroke: A safe and humane screening tool. *Age and Ageing*, *29*(6), 495–499. doi:10.1093/ageing/29.6.495 PMID:11191240

Sommerfeld, D. K., Gripenstedt, U., & Welmer, A.-K. (2012). Spasticity after stroke: An overview of prevalence, test instruments, and treatments. *American Journal of Physical Medicine & Rehabilitation*, *91*(9), 814–820. doi:10.1097/PHM.0b013e31825f13a3 PMID:22760104

Sorensen, A. G., & Ay, H. (2011). Transient ischemic attack: Definition, diagnosis, and risk stratification. *Neuroimaging Clinics of North America*, *21*(2), 303–313. doi:10.1016/j.nic.2011.01.013 PMID:21640301

Souzaa, D. C. B., Santosb, M. S., Ribeirob, N. M. S., & Maldonado, I. L. (2019). Inpatient trunk exercises after recent stroke: An update meta-analysis of randomized controlled trials. *NeuroRehabilitation, 44*(3), 369–377. doi:10.3233/NRE-182585

Stone, S., Wilson, B., Wroot, A., Halligan, P., Lange, L., Marshall, J., & Greenwood, R. (1991). The assessment of visuo-spatial neglect after acute stroke. *Journal of Neurology, Neurosurgery, and Psychiatry*, *54*(4), 345–350. doi:10.1136/jnnp.54.4.345 PMID:2056321

Sun, J.-H., Tan, L., & Yu, J.-T. (2014). Post-stroke cognitive impairment: Epidemiology, mechanisms and management. *Annals of Translational Medicine*, *2*(8). Advance online publication. doi:10.3978/j.issn.2305-5839.2014.08.05 PMID:25333055

Sura, L., Madhavan, A., Carnaby, G., & Crary, M. A. (2012). Dysphagia in the elderly: Management and nutritional considerations. *Clinical Interventions in Aging*, *7*, 287. doi:10.2147/CIA.S23404 PMID:22956864

Tanaka, S., Hachisuka, K., & Ogata, H. (1997). Trunk rotatory muscle performance in post-stroke hemiplegic patients1. *American Journal of Physical Medicine & Rehabilitation*, *76*(5), 366–369. doi:10.1097/00002060-199709000-00003 PMID:9354489

Tanaka, S., Hachisuka, K., & Ogata, H. (1998). Muscle strength of trunk flexion-extension in post-stroke hemiplegic patients. *American Journal of Physical Medicine & Rehabilitation*, *77*(4), 288–290. doi:10.1097/00002060-199807000-00005 PMID:9715916

Teasell, R., Foley, N., Pereira, S., Sequeira, K., & Miller, T. (2012). Evidence to practice: Botulinum toxin in the treatment of spasticity post stroke. *Topics in Stroke Rehabilitation*, *19*(2), 115–121. doi:10.1310/tsr1902-115 PMID:22436359

Teichroew, J. K. (2016). *Chronic Diseases: An Encyclopedia of Causes, Effects, and Treatments* (2 vols.). ABC-CLIO.

Tekin Dal. (2019). Baş Dönmesi Olan Hastaların Vestibüler Rehabilitasyonunda Aktivite Eğitiminin Etkisi. Academic Press.

Telian, S. A. (1991). Physical therapy program for vestibular rehabilitation. *The American Journal of Otology*, *12*(3). PMID:1882973

Terre, R., & Mearin, F. (2006). Oropharyngeal dysphagia after the acute phase of stroke: Predictors of aspiration. *Neurogastroenterology and Motility*, *18*(3), 200–205. doi:10.1111/j.1365-2982.2005.00729.x PMID:16487410

Theberge, A., Desrosiers, C., Descoteaux, M., & Jodoin, P.-M. (2020). Track-To-Learn: A general framework for tractography with deep reinforcement learning. doi:10.1101/2020.11.16.385229

Thibaut, A., Chatelle, C., Ziegler, E., Bruno, M.-A., Laureys, S., & Gosseries, O. (2013). Spasticity after stroke: Physiology, assessment and treatment. *Brain Injury: [BI]*, *27*(10), 1093–1105. doi:10.3109/02699052.2013.804202 PMID:23885710

Tsuji, T., Liu, M., Sonoda, S., Domen, K., & Chino, N. (2000). The stroke impairment assessment set: Its internal consistency and predictive validity. *Archives of Physical Medicine and Rehabilitation*, *81*(7), 863–868. doi:10.1053/apmr.2000.6275 PMID:10895996

Turner-Stokes, L., & Jackson, D. (2002). Shoulder pain after stroke: A review of the evidence base to inform the development of an integrated care pathway. *Clinical Rehabilitation*, *16*(3), 276–298. doi:10.1191/0269215502cr491oa PMID:12017515

Ukraintseva, S., Sloan, F., Arbeev, K., & Yashin, A. (2006). Increasing rates of dementia at time of declining mortality from stroke. *Stroke*, *37*(5), 1155–1159. doi:10.1161/01.STR.0000217971.88034.e9 PMID:16601210

Vafadar, A. K., Côté, J. N., & Archambault, P. S. (2015). Effectiveness of functional electrical stimulation in improving clinical outcomes in the upper arm following stroke: A systematic review and meta-analysis. *BioMed Research International*, *2015*, 1–14. Advance online publication. doi:10.1155/2015/729768 PMID:25685805

Van der Burg, J. C., Pijnappels, M., & van Dieen, J. H. (2005). Out of plane trunk movements and trunk muscle activity after a trip during walking. *Experimental Brain Research*, *165*(3), 407–412. doi:10.100700221-005-2312-z PMID:15875169

Van Ouwenaller, C., Laplace, P., & Chantraine, A. (1986). Painful shoulder in hemiplegia. *Archives of Physical Medicine and Rehabilitation*, *67*(1), 23–26. PMID:3942479

Van Peppen, R. P., Kwakkel, G., Wood-Dauphinee, S., Hendriks, H. J., Van der Wees, P. J., & Dekker, J. (2004). The impact of physical therapy on functional outcomes after stroke: What's the evidence? *Clinical Rehabilitation*, *18*(8), 833–862. doi:10.1191/0269215504cr843oa PMID:15609840

VanCriekinge, T., Saeys, W., Hallemans, A., Velghe, S., Viskens, P. J., Vereeck, L., De Hertogh, W., & Truijen, S. (2017). Trunk biomechanics during hemiplegic gait after stroke: A systematic review. *Gait & Posture*, *54*, 133–143. doi:10.1016/j.gaitpost.2017.03.004 PMID:28288334

Verheyden, G., Nieuwboer, A., Mertin, J., Preger, R., Kiekens, C., & De Weerdt, W. (2004). The trunk impairment scale: A new tool to measure motor impairment of the trunk after stroke. *Clinical Rehabilitation*, *18*(3), 326–334. doi:10.1191/0269215504cr733oa PMID:15137564

Verheyden, G., Nieuwboer, A., Van de Winckel, A., & De Weerdt, W. (2007). Clinical tools to measure trunk performance after stroke: A systematic review of the literatüre. *Clinical Rehabilitation*, *21*(5), 387–394. doi:10.1177/0269215507074055 PMID:17613559

Verma, A., Cairns, J. A., Mitchell, L. B., Macle, L., Stiell, I. G., Gladstone, D., McMurtry, M. S., Connolly, S., Cox, J. L., Dorian, P., Ivers, N., Leblanc, K., Nattel, S., & Healey, J. S. (2014). 2014 focused update of the Canadian Cardiovascular Society Guidelines for the management of atrial fibrillation. *The Canadian Journal of Cardiology*, *30*(10), 1114–1130. doi:10.1016/j.cjca.2014.08.001 PMID:25262857

Villa, R. F., Ferrari, F., & Moretti, A. (2018). Post-stroke depression: Mechanisms and pharmacological treatment. *Pharmacology & Therapeutics*, *184*, 131–144. doi:10.1016/j.pharmthera.2017.11.005 PMID:29128343

Wade, D. T., & Hewer, R. L. (1987). Motor loss and swallowing difficulty after stroke: Frequency, recovery and prognosis. *Acta Neurologica Scandinavica*, *76*(1), 50–54. doi:10.1111/j.1600-0404.1987.tb03543.x PMID:3630644

Wade, D. T., Skilbeck, C. E., & Hewer, R. L. (1983). Predicting Barthel ADL score at 6 months after an acute stroke. *Archives of Physical Medicine and Rehabilitation*, *64*(1), 24–28. PMID:6849630

Wajngarten, M., & Silva, G. S. (2019). Hypertension and Stroke: Update on Treatment. *European Cardiology*, *14*(2), 111–115. doi:10.15420/ecr.2019.11.1 PMID:31360232

Walsh, K. (2001). Management of shoulder pain in patients with stroke. *Postgraduate Medical Journal*, *77*(912), 645–649. doi:10.1136/pmj.77.912.645 PMID:11571371

Wang, Y., Pan, Y., Zhao, X., Li, H., Wang, D., Johnston, S. C., Liu, L., Meng, X., Wang, A., Wang, C., & Wang, C. (2015). Clopidogrel With Aspirin in Acute Minor Stroke or Transient Ischemic Attack (CHANCE) Trial: One-Year Outcomes. *Circulation*, *132*(1), 40–46. doi:10.1161/CIRCULATIONAHA.114.014791 PMID:25957224

Wardlaw, J. M., Chappell, F. M., Stevenson, M., De Nigris, E., Thomas, S., Gillard, J., Berry, E., Young, G., Rothwell, P., Roditi, G., Gough, M., Brennan, A., Bamford, J., & Best, J. (2006). Accurate, practical and cost-effective assessment of carotid stenosis in the UK. *Health Technology Assessment*, *10*(30), 1–182. doi:10.3310/hta10300 PMID:16904049

Yang, S. N. (2017). Current Evidence for Post Stroke Aphasia Treatment. *Brain & Neurorehabilitation*, *10*(2), e15. Advance online publication. doi:10.12786/bn.2017.10.e15

Yılmaz, Ö. (2004). Vestibüler rehabilitasyon. *Ankara Üniveristesi Sağlık Hizmetleri Meslek Yüksekokulu Dergisi*, *5*(1), 11–15.

Yu, D. (2004). Shoulder pain in hemiplegia. *Physical Medicine and Rehabilitation Clinics of North America*, *15*(3), 683–697. doi:10.1016/S1047-9651(03)00130-X PMID:15219895

Zhang, X., Shu, B., Zhang, D., Huang, L., Fu, Q., & Du, G. (2018). The efficacy and safety of pharmacological treatments for post-stroke aphasia. *CNS & Neurological Disorders-Drug Targets, 17*(7), 509-521. doi:10.2174/1871527317666180706143051

Zhang, J. T., Chen, K. P., & Zhang, S. (2015). Efficacy and safety of oral anticoagulants versus aspirin for patients with atrial fibrillation: A meta-analysis. *Medicine, 94*(4), 0000000000000409. doi:10.1097/MD.0000000000000409 PMID:25634169

Zhang, Q., Fu, C., Liang, Z., Peng, L., Xiong, F., Chen, L., He, C., & Wei, Q. (2020). The effect of adding trunk restraint to task-oriented training in improving function in stroke patients: A systematic review and meta-analysis. *NeuroRehabilitation, 46*(1), 95–108. doi:10.3233/NRE-192964 PMID:32039877

Zorowitz, R. D., & Tietjen, G. E. (1999). Medical complications after stroke. *Journal of Stroke and Cerebrovascular Diseases, 8*(3), 192–196. doi:10.1016/S1052-3057(99)80026-3 PMID:17895163

# Related References

To continue our tradition of advancing information science and technology research, we have compiled a list of recommended IGI Global readings. These references will provide additional information and guidance to further enrich your knowledge and assist you with your own research and future publications.

Abu Seman, S. A., & Ramayah, T. (2017). Are We Ready to App?: A Study on mHealth Apps, Its Future, and Trends in Malaysia Context. In J. Pelet (Ed.), *Mobile Platforms, Design, and Apps for Social Commerce* (pp. 69–83). Hershey, PA: IGI Global. doi:10.4018/978-1-5225-2469-4.ch005

Aggarwal, S., & Azad, V. (2017). A Hybrid System Based on FMM and MLP to Diagnose Heart Disease. In S. Bhattacharyya, S. De, I. Pan, & P. Dutta (Eds.), *Intelligent Multidimensional Data Clustering and Analysis* (pp. 293–325). Hershey, PA: IGI Global. doi:10.4018/978-1-5225-1776-4.ch011

Akaichi, J., & Mhadhbi, L. (2016). A Clinical Decision Support System: Ontology-Driven Approach for Effective Emergency Management. In J. Moon & M. Galea (Eds.), *Improving Health Management through Clinical Decision Support Systems* (pp. 270–294). Hershey, PA: IGI Global. doi:10.4018/978-1-4666-9432-3.ch013

Akerkar, R. (2016). Towards an Intelligent Integrated Approach for Clinical Decision Support. In A. Aggarwal (Ed.), *Managing Big Data Integration in the Public Sector* (pp. 187–205). Hershey, PA: IGI Global. doi:10.4018/978-1-4666-9649-5.ch011

Al-Busaidi, S. S. (2018). Interdisciplinary Relationships Between Medicine and Social Sciences. In M. Al-Suqri, A. Al-Kindi, S. AlKindi, & N. Saleem (Eds.), *Promoting Interdisciplinarity in Knowledge Generation and Problem Solving* (pp. 124–137). Hershey, PA: IGI Global. doi:10.4018/978-1-5225-3878-3.ch009

Al Kareh, T., & Thoumy, M. (2018). The Impact of Health Information Digitization on the Physiotherapist-Patient Relationship: A Pilot Study of the Lebanese Community. *International Journal of Healthcare Information Systems and Informatics, 13*(2), 29–53. doi:10.4018/IJHISI.2018040103

Alenzuela, R. (2017). Research, Leadership, and Resource-Sharing Initiatives: The Role of Local Library Consortia in Access to Medical Information. In S. Ram (Ed.), *Library and Information Services for Bioinformatics Education and Research* (pp. 199–211). Hershey, PA: IGI Global. doi:10.4018/978-1-5225-1871-6.ch012

Ameri, H., Alizadeh, S., & Noughabi, E. A. (2017). Application of Data Mining Techniques in Clinical Decision Making: A Literature Review and Classification. In E. Noughabi, B. Raahemi, A. Albadvi, & B. Far (Eds.), *Handbook of Research on Data Science for Effective Healthcare Practice and Administration* (pp. 257–295). Hershey, PA: IGI Global. doi:10.4018/978-1-5225-2515-8.ch012

Anand, S. (2017). Medical Image Enhancement Using Edge Information-Based Methods. In B. Singh (Ed.), *Computational Tools and Techniques for Biomedical Signal Processing* (pp. 123–148). Hershey, PA: IGI Global. doi:10.4018/978-1-5225-0660-7.ch006

Angjellari-Dajci, F., Sapienza, C., Lawless, W. F., & Kavanagh, K. (2016). Human Capital Accumulation in Medical Simulated Learning Environments: A Framework for Economic Evaluation. In M. Russ (Ed.), *Quantitative Multidisciplinary Approaches in Human Capital and Asset Management* (pp. 65–88). Hershey, PA: IGI Global. doi:10.4018/978-1-4666-9652-5.ch004

Anil, M., Ayyildiz-Tamis, D., Tasdemir, S., Sendemir-Urkmez, A., & Gulce-Iz, S. (2016). Bioinspired Materials and Biocompatibility. In M. Bououdina (Ed.), *Emerging Research on Bioinspired Materials Engineering* (pp. 294–322). Hershey, PA: IGI Global. doi:10.4018/978-1-4666-9811-6.ch011

Anton, J. L., Soriano, J. V., Martinez, M. I., & Garcia, F. B. (2018). Comprehensive E-Learning Appraisal System. In M. Khosrow-Pour, D.B.A. (Ed.), Encyclopedia of Information Science and Technology, Fourth Edition (pp. 5787-5799). Hershey, PA: IGI Global. doi:10.4018/978-1-5225-2255-3.ch503

Arabi, M. (2016). Redesigning the Organizational Structure to Reach Efficiency: The Case of Ministry of Health and Medical Education of Iran. In R. Gholipour & K. Rouzbehani (Eds.), *Social, Economic, and Political Perspectives on Public Health Policy-Making* (pp. 257–294). Hershey, PA: IGI Global. doi:10.4018/978-1-4666-9944-1.ch012

Assis-Hassid, S., Heart, T., Reychav, I., & Pliskin, J. S. (2016). Modelling Factors Affecting Patient-Doctor-Computer Communication in Primary Care. *International Journal of Reliable and Quality E-Healthcare, 5*(1), 1–17. doi:10.4018/IJRQEH.2016010101

Atzori, B., Hoffman, H. G., Vagnoli, L., Messeri, A., & Grotto, R. L. (2018). Virtual Reality as Distraction Technique for Pain Management in Children and Adolescents. In M. Khosrow-Pour, D.B.A. (Ed.), Encyclopedia of Information Science and Technology, Fourth Edition (pp. 5955-5965). Hershey, PA: IGI Global. doi:10.4018/978-1-5225-2255-3.ch518

Audibert, M., Mathonnat, J., Pélissier, A., & Huang, X. X. (2018). The Impact of the New Rural Cooperative Medical Scheme on Township Hospitals' Utilization and Income Structure in Weifang Prefecture, China. In I. Management Association (Ed.), Health Economics and Healthcare Reform: Breakthroughs in Research and Practice (pp. 109-121). Hershey, PA: IGI Global. doi:10.4018/978-1-5225-3168-5.ch007

Ayyachamy, S. (2016). Analysis and Comparison of Developed 2D Medical Image Database Design using Registration Scheme, Retrieval Scheme, and Bag-of-Visual-Words. In N. Dey & A. Ashour (Eds.), *Classification and Clustering in Biomedical Signal Processing* (pp. 149–168). Hershey, PA: IGI Global. doi:10.4018/978-1-5225-0140-4.ch007

Babalola, A. C. (2016). Literacy and Decision Making on Health Issues among Married Women in Southwest Nigeria. In V. Wang (Ed.), *Handbook of Research on Advancing Health Education through Technology* (pp. 191–212). Hershey, PA: IGI Global. doi:10.4018/978-1-4666-9494-1.ch009

Babu, A., & Ayyappan, S. (2017). A Methodological Evaluation of Crypto-Watermarking System for Medical Images. In C. Bhatt & S. Peddoju (Eds.), *Cloud Computing Systems and Applications in Healthcare* (pp. 189–217). Hershey, PA: IGI Global. doi:10.4018/978-1-5225-1002-4.ch010

Bakke, A. (2017). Ethos in E-Health: From Informational to Interactive Websites. In M. Folk & S. Apostel (Eds.), *Establishing and Evaluating Digital Ethos and Online Credibility* (pp. 85–103). Hershey, PA: IGI Global. doi:10.4018/978-1-5225-1072-7.ch005

Bali, S. (2018). Enhancing the Reach of Health Care Through Telemedicine: Status and New Possibilities in Developing Countries. In U. Pandey & V. Indrakanti (Eds.), *Open and Distance Learning Initiatives for Sustainable Development* (pp. 339–354). Hershey, PA: IGI Global. doi:10.4018/978-1-5225-2621-6.ch019

Barrett, T. E. (2018). Essentials for Education and Training for Tomorrow's Physicians. In C. Smith (Ed.), *Exploring the Pressures of Medical Education From a Mental Health and Wellness Perspective* (pp. 230–252). Hershey, PA: IGI Global. doi:10.4018/978-1-5225-2811-1.ch010

Barrett, T. J. (2016). Knowledge in Action: Fostering Health Education through Technology. In V. Wang (Ed.), *Handbook of Research on Advancing Health Education through Technology* (pp. 39–62). Hershey, PA: IGI Global. doi:10.4018/978-1-4666-9494-1.ch003

Berrahal, S., & Boudriga, N. (2017). The Risks of Wearable Technologies to Individuals and Organizations. In A. Marrington, D. Kerr, & J. Gammack (Eds.), *Managing Security Issues and the Hidden Dangers of Wearable Technologies* (pp. 18–46). Hershey, PA: IGI Global. doi:10.4018/978-1-5225-1016-1.ch002

Bhagat, A. P., & Atique, M. (2016). Medical Image Mining Using Fuzzy Connectedness Image Segmentation: Efficient Retrieval of Patients' Stored Images. In W. Karâa & N. Dey (Eds.), *Biomedical Image Analysis and Mining Techniques for Improved Health Outcomes* (pp. 184–209). Hershey, PA: IGI Global. doi:10.4018/978-1-4666-8811-7.ch009

Bhargava, P. (2018). Blended Learning: An Effective Application to Clinical Teaching. In S. Tang & C. Lim (Eds.), *Preparing the Next Generation of Teachers for 21st Century Education* (pp. 302–320). Hershey, PA: IGI Global. doi:10.4018/978-1-5225-4080-9.ch018

Binh, N. T., & Tuyet, V. T. (2016). The Combination of Adaptive Filters to Improve the Quality of Medical Images in New Wavelet Domain. In N. Dey & A. Ashour (Eds.), *Classification and Clustering in Biomedical Signal Processing* (pp. 46–76). Hershey, PA: IGI Global. doi:10.4018/978-1-5225-0140-4.ch003

Bird, J. L. (2017). Writing Healing Narratives. In V. Bryan & J. Bird (Eds.), *Healthcare Community Synergism between Patients, Practitioners, and Researchers* (pp. 1–28). Hershey, PA: IGI Global. doi:10.4018/978-1-5225-0640-9.ch001

Bottrighi, A., Leonardi, G., Piovesan, L., & Terenziani, P. (2016). Knowledge-Based Support to the Treatment of Exceptions in Computer Interpretable Clinical Guidelines. *International Journal of Knowledge-Based Organizations*, *6*(3), 1–27. doi:10.4018/IJKBO.2016070101

Bougoulias, K., Kouris, I., Prasinos, M., Giokas, K., & Koutsouris, D. (2017). Ob/Gyn EMR Software: A Solution for Obstetricians and Gynecologists. In A. Moumtzoglou (Ed.), *Design, Development, and Integration of Reliable Electronic Healthcare Platforms* (pp. 101–111). Hershey, PA: IGI Global. doi:10.4018/978-1-5225-1724-5.ch006

Bouslimi, R., Ayadi, M. G., & Akaichi, J. (2018). Medical Image Retrieval in Healthcare Social Networks. *International Journal of Healthcare Information Systems and Informatics*, *13*(2), 13–28. doi:10.4018/IJHISI.2018040102

Bouzaabia, O., Bouzaabia, R., & Mejri, K. (2017). Role of Internet in the Development of Medical Tourism Service in Tunisia. In A. Capatina & E. Rancati (Eds.), *Key Challenges and Opportunities in Web Entrepreneurship* (pp. 211–241). Hershey, PA: IGI Global. doi:10.4018/978-1-5225-2466-3.ch009

Briz-Ponce, L., Juanes-Méndez, J. A., & García-Peñalvo, F. J. (2016). The Role of Gender in Technology Acceptance for Medical Education. In M. Cruz-Cunha, I. Miranda, R. Martinho, & R. Rijo (Eds.), *Encyclopedia of E-Health and Telemedicine* (pp. 1013–1027). Hershey, PA: IGI Global. doi:10.4018/978-1-4666-9978-6.ch079

Briz-Ponce, L., Juanes-Méndez, J. A., & García-Peñalvo, F. J. (2016). A Research Contribution to the Analysis of Mobile Devices in Higher Education from Medical Students' Point of View. In L. Briz-Ponce, J. Juanes-Méndez, & F. García-Peñalvo (Eds.), *Handbook of Research on Mobile Devices and Applications in Higher Education Settings* (pp. 196–221). Hershey, PA: IGI Global. doi:10.4018/978-1-5225-0256-2.ch009

Brown-Jackson, K. L. (2018). Telemedicine and Telehealth: Academics Engaging the Community in a Call to Action. In S. Burton (Ed.), *Engaged Scholarship and Civic Responsibility in Higher Education* (pp. 166–193). Hershey, PA: IGI Global. doi:10.4018/978-1-5225-3649-9.ch008

Bwalya, K. J. (2017). Next Wave of Tele-Medicine: Virtual Presence of Medical Personnel. In K. Moahi, K. Bwalya, & P. Sebina (Eds.), *Health Information Systems and the Advancement of Medical Practice in Developing Countries* (pp. 168–180). Hershey, PA: IGI Global. doi:10.4018/978-1-5225-2262-1.ch010

Carlton, E. L., Holsinger, J. W. Jr, & Anunobi, N. (2016). Physician Engagement with Health Information Technology: Implications for Practice and Professionalism. *International Journal of Computers in Clinical Practice*, *1*(2), 51–73. doi:10.4018/IJCCP.2016070103

Castillo-Page, L., Eliason, J., Conrad, S. S., & Nivet, M. A. (2016). Diversity in Undergraduate Medical Education: An Examination of Organizational Culture and Climate in Medical Schools. In C. Scott & J. Sims (Eds.), *Developing Workforce Diversity Programs, Curriculum, and Degrees in Higher Education* (pp. 304–326). Hershey, PA: IGI Global. doi:10.4018/978-1-5225-0209-8.ch016

Cebeci, H. I., & Hiziroglu, A. (2016). Review of Business Intelligence and Intelligent Systems in Healthcare Domain. In N. Celebi (Ed.), *Intelligent Techniques for Data Analysis in Diverse Settings* (pp. 192–206). Hershey, PA: IGI Global. doi:10.4018/978-1-5225-0075-9.ch009

Celik, G. (2018). Determining Headache Diseases With Genetic Algorithm. In U. Kose, G. Guraksin, & O. Deperlioglu (Eds.), *Nature-Inspired Intelligent Techniques for Solving Biomedical Engineering Problems* (pp. 249–262). Hershey, PA: IGI Global. doi:10.4018/978-1-5225-4769-3.ch012

Chakraborty, C., Gupta, B., & Ghosh, S. K. (2016). Mobile Telemedicine Systems for Remote Patient's Chronic Wound Monitoring. In A. Moumtzoglou (Ed.), *M-Health Innovations for Patient-Centered Care* (pp. 213–239). Hershey, PA: IGI Global. doi:10.4018/978-1-4666-9861-1.ch011

Chakraborty, S., Chatterjee, S., Ashour, A. S., Mali, K., & Dey, N. (2018). Intelligent Computing in Medical Imaging: A Study. In N. Dey (Ed.), *Advancements in Applied Metaheuristic Computing* (pp. 143–163). Hershey, PA: IGI Global. doi:10.4018/978-1-5225-4151-6.ch006

Chavez, A., & Kovarik, C. (2017). Open Source Technology for Medical Practice in Developing Countries. In K. Moahi, K. Bwalya, & P. Sebina (Eds.), *Health Information Systems and the Advancement of Medical Practice in Developing Countries* (pp. 33–59). Hershey, PA: IGI Global. doi:10.4018/978-1-5225-2262-1.ch003

Chen, E. T. (2016). Examining the Influence of Information Technology on Modern Health Care. In P. Manolitzas, E. Grigoroudis, N. Matsatsinis, & D. Yannacopoulos (Eds.), *Effective Methods for Modern Healthcare Service Quality and Evaluation* (pp. 110–136). Hershey, PA: IGI Global. doi:10.4018/978-1-4666-9961-8.ch006

Chin, S. (2016). Surviving Sandy: Recovering Collections after a Natural Disaster. In E. Decker & J. Townes (Eds.), *Handbook of Research on Disaster Management and Contingency Planning in Modern Libraries* (pp. 366–388). Hershey, PA: IGI Global. doi:10.4018/978-1-4666-8624-3.ch016

Chojnacki, M., & Wójcik, A. (2016). Security and the Role of New Technologies and Innovation in Medical Ethics. In A. Rosiek & K. Leksowski (Eds.), *Organizational Culture and Ethics in Modern Medicine* (pp. 52–77). Hershey, PA: IGI Global. doi:10.4018/978-1-4666-9658-7.ch003

Cılız, N., Yıldırım, H., & Temizel, Ş. (2016). Structure Development for Effective Medical Waste and Hazardous Waste Management System. In U. Akkucuk (Ed.), *Handbook of Research on Waste Management Techniques for Sustainability* (pp. 303–327). Hershey, PA: IGI Global. doi:10.4018/978-1-4666-9723-2.ch016

Ciufudean, C. (2018). Innovative Formalism for Biological Data Analysis. In M. Khosrow-Pour, D.B.A. (Ed.), Encyclopedia of Information Science and Technology, Fourth Edition (pp. 1814-1824). Hershey, PA: IGI Global. doi:10.4018/978-1-5225-2255-3.ch158

Ciufudean, C., & Ciufudean, O. (2016). Tele-Monitoring for Medical Diagnosis Availability. In M. Cruz-Cunha, I. Miranda, R. Martinho, & R. Rijo (Eds.), *Encyclopedia of E-Health and Telemedicine* (pp. 401–411). Hershey, PA: IGI Global. doi:10.4018/978-1-4666-9978-6.ch032

Colaguori, R., & Danesi, M. (2017). Medical Semiotics: A Revisitation and an Exhortation. *International Journal of Semiotics and Visual Rhetoric*, *1*(1), 11–18. doi:10.4018/IJSVR.2017010102

Cole, A. W., & Salek, T. A. (2017). Adopting a Parasocial Connection to Overcome Professional Kakoethos in Online Health Information. In M. Folk & S. Apostel (Eds.), *Establishing and Evaluating Digital Ethos and Online Credibility* (pp. 104–120). Hershey, PA: IGI Global. doi:10.4018/978-1-5225-1072-7.ch006

Contreras, E. C., & Puente, G. J. (2016). How to Identify Rheumatic Diseases by General Physicians. In T. Gasmelseid (Ed.), *Advancing Pharmaceutical Processes and Tools for Improved Health Outcomes* (pp. 136–166). Hershey, PA: IGI Global. doi:10.4018/978-1-5225-0248-7.ch006

D'Andrea, A., Ferri, F., & Grifoni, P. (2016). RFID Technologies in Healthcare Setting: Applications and Future Perspectives. *International Journal of Computers in Clinical Practice*, *1*(1), 15–27. doi:10.4018/IJCCP.2016010102

Damianakis, A., Kallonis, P., Loudos, G., Tsatsos, D., & Tsoukalis, A. (2016). Exploiting 3D Medical Equipment Simulations to Support Biomedical Engineering Academic Courses: Design Methodology and Implementation in a Small Scale National Project. In B. Khan (Ed.), *Revolutionizing Modern Education through Meaningful E-Learning Implementation* (pp. 277–295). Hershey, PA: IGI Global. doi:10.4018/978-1-5225-0466-5.ch015

Demiroz, E. (2016). Principles of Instructional Design for E-Learning and Online Learning Practices: Implications for Medical Education. In V. Wang (Ed.), *Handbook of Research on Advancing Health Education through Technology* (pp. 419–451). Hershey, PA: IGI Global. doi:10.4018/978-1-4666-9494-1.ch018

Deperlioglu, O. (2018). Intelligent Techniques Inspired by Nature and Used in Biomedical Engineering. In U. Kose, G. Guraksin, & O. Deperlioglu (Eds.), *Nature-Inspired Intelligent Techniques for Solving Biomedical Engineering Problems* (pp. 51–77). Hershey, PA: IGI Global. doi:10.4018/978-1-5225-4769-3.ch003

Dey, N., & Ashour, A. S. (2018). Meta-Heuristic Algorithms in Medical Image Segmentation: A Review. In N. Dey (Ed.), *Advancements in Applied Metaheuristic Computing* (pp. 185–203). Hershey, PA: IGI Global. doi:10.4018/978-1-5225-4151-6.ch008

Dey, N., Ashour, A. S., & Althoupety, A. S. (2017). Thermal Imaging in Medical Science. In V. Santhi (Ed.), *Recent Advances in Applied Thermal Imaging for Industrial Applications* (pp. 87–117). Hershey, PA: IGI Global. doi:10.4018/978-1-5225-2423-6.ch004

Di Virgilio, F., Camillo, A. A., & Camillo, I. C. (2017). The Impact of Social Network on Italian Users' Behavioural Intention for the Choice of a Medical Tourist Destination. *International Journal of Tourism and Hospitality Management in the Digital Age*, *1*(1), 36–49. doi:10.4018/IJTHMDA.2017010103

Dias, C. M., Ribeiro, A. G., & Furtado, S. F. (2016). An Overview about the Use of Healthcare Applications on Mobile Devices. In M. Cruz-Cunha, I. Miranda, R. Martinho, & R. Rijo (Eds.), *Encyclopedia of E-Health and Telemedicine* (pp. 285–298). Hershey, PA: IGI Global. doi:10.4018/978-1-4666-9978-6.ch024

Dogra, A. K., & Dogra, P. (2017). The Medical Tourism Industry in the BRIC Nations: An Indian Analysis. In M. Dhiman (Ed.), *Opportunities and Challenges for Tourism and Hospitality in the BRIC Nations* (pp. 320–336). Hershey, PA: IGI Global. doi:10.4018/978-1-5225-0708-6.ch020

Drowos, J. L., & Wood, S. K. (2017). Preparing Future Physicians to Adapt to the Changing Health Care System: Promoting Humanism through Curricular Design. In V. Bryan & J. Bird (Eds.), *Healthcare Community Synergism between Patients, Practitioners, and Researchers* (pp. 106–125). Hershey, PA: IGI Global. doi:10.4018/978-1-5225-0640-9.ch006

Dulam, K. (2017). Medical Patents and Impact on Availability and Affordability of Essential Medicines in India. In R. Aggarwal & R. Kaur (Eds.), *Patent Law and Intellectual Property in the Medical Field* (pp. 41–57). Hershey, PA: IGI Global. doi:10.4018/978-1-5225-2414-4.ch003

Dutta, P. (2017). Decision Making in Medical Diagnosis via Distance Measures on Interval Valued Fuzzy Sets. *International Journal of System Dynamics Applications*, *6*(4), 63–83. doi:10.4018/IJSDA.2017100104

Edoh, T. O., Pawar, P. A., Brügge, B., & Teege, G. (2016). A Multidisciplinary Remote Healthcare Delivery System to Increase Health Care Access, Pathology Screening, and Treatment in Developing Countries: The Case of Benin. *International Journal of Healthcare Information Systems and Informatics*, *11*(4), 1–31. doi:10.4018/IJHISI.2016100101

El Guemhioui, K., & Demurjian, S. A. (2017). Semantic Reconciliation of Electronic Health Records Using Semantic Web Technologies. *International Journal of Information Technology and Web Engineering*, *12*(2), 26–48. doi:10.4018/IJITWE.2017040102

Ellouze, N., Rekhis, S., & Boudriga, N. (2016). Forensic Investigation of Digital Crimes in Healthcare Applications. In O. Isafiade & A. Bagula (Eds.), *Data Mining Trends and Applications in Criminal Science and Investigations* (pp. 169–210). Hershey, PA: IGI Global. doi:10.4018/978-1-5225-0463-4.ch007

Eneanya, A. N. (2016). Health Policy Implementation and Its Barriers: The Case Study of US Health System. In R. Gholipour & K. Rouzbehani (Eds.), *Social, Economic, and Political Perspectives on Public Health Policy-Making* (pp. 42–63). Hershey, PA: IGI Global. doi:10.4018/978-1-4666-9944-1.ch003

Entico, G. J. (2016). Knowledge Management and the Medical Health Librarians: A Perception Study. In J. Yap, M. Perez, M. Ayson, & G. Entico (Eds.), *Special Library Administration, Standardization and Technological Integration* (pp. 52–77). Hershey, PA: IGI Global. doi:10.4018/978-1-4666-9542-9.ch003

Ferradji, M. A., & Zidani, A. (2016). Collaborative Environment for Remote Clinical Reasoning Learning. *International Journal of E-Health and Medical Communications*, *7*(4), 62–81. doi:10.4018/IJEHMC.2016100104

Fisher, J. (2018). Sociological Perspectives on Improving Medical Diagnosis Emphasizing CAD. In M. Khosrow-Pour, D.B.A. (Ed.), Encyclopedia of Information Science and Technology, Fourth Edition (pp. 1017-1024). Hershey, PA: IGI Global. doi:10.4018/978-1-5225-2255-3.ch088

Flores, C. D., Respício, A., Coelho, H., Bez, M. R., & Fonseca, J. M. (2016). Simulation for Medical Training. In M. Cruz-Cunha, I. Miranda, R. Martinho, & R. Rijo (Eds.), *Encyclopedia of E-Health and Telemedicine* (pp. 827–842). Hershey, PA: IGI Global. doi:10.4018/978-1-4666-9978-6.ch064

Frank, E. M. (2018). Healthcare Education: Integrating Simulation Technologies. In V. Bryan, A. Musgrove, & J. Powers (Eds.), *Handbook of Research on Human Development in the Digital Age* (pp. 163–182). Hershey, PA: IGI Global. doi:10.4018/978-1-5225-2838-8.ch008

Garner, G. (2018). Foundations for Yoga Practice in Rehabilitation. In S. Telles & N. Singh (Eds.), *Research-Based Perspectives on the Psychophysiology of Yoga* (pp. 263–307). Hershey, PA: IGI Global. doi:10.4018/978-1-5225-2788-6.ch015

Gasmelseid, T. M. (2016). Improving Pharmaceutical Care through the Use of Intelligent Pharmacoinformatics. In T. Gasmelseid (Ed.), *Advancing Pharmaceutical Processes and Tools for Improved Health Outcomes* (pp. 167–188). Hershey, PA: IGI Global. doi:10.4018/978-1-5225-0248-7.ch007

Gavurová, B., Kováč, V., & Šoltés, M. (2018). Medical Equipment and Economic Determinants of Its Structure and Regulation in the Slovak Republic. In M. Khosrow-Pour, D.B.A. (Ed.), Encyclopedia of Information Science and Technology, Fourth Edition (pp. 5841-5852). Hershey, PA: IGI Global. doi:10.4018/978-1-5225-2255-3.ch508

Ge, X., Wang, Q., Huang, K., Law, V., & Thomas, D. C. (2017). Designing Simulated Learning Environments and Facilitating Authentic Learning Experiences in Medical Education. In J. Stefaniak (Ed.), *Advancing Medical Education Through Strategic Instructional Design* (pp. 77–100). Hershey, PA: IGI Global. doi:10.4018/978-1-5225-2098-6.ch004

Gewald, H., & Gewald, C. (2018). Inhibitors of Physicians' Use of Mandatory Hospital Information Systems (HIS). *International Journal of Healthcare Information Systems and Informatics, 13*(1), 29–44. doi:10.4018/IJHISI.2018010103

Ghosh, D., & Dinda, S. (2017). Health Infrastructure and Economic Development in India. In R. Das (Ed.), *Social, Health, and Environmental Infrastructures for Economic Growth* (pp. 99–119). Hershey, PA: IGI Global. doi:10.4018/978-1-5225-2364-2.ch006

Ghosh, K., & Sen, K. C. (2017). The Potential of Crowdsourcing in the Health Care Industry. In N. Wickramasinghe (Ed.), *Handbook of Research on Healthcare Administration and Management* (pp. 418–427). Hershey, PA: IGI Global. doi:10.4018/978-1-5225-0920-2.ch024

Ghrab, S., & Saad, I. (2016). Identifying Crucial Know-How and Knowing-That for Medical Decision Support. *International Journal of Decision Support System Technology, 8*(4), 14–33. doi:10.4018/IJDSST.2016100102

Giokas, K., Tsirmpas, C., Anastasiou, A., Iliopoulou, D., Costarides, V., & Koutsouris, D. (2016). Contemporary Heart Failure Treatment Based on Improved Knowledge and Personalized Care of Comorbidities. In D. Fotiadis (Ed.), *Handbook of Research on Trends in the Diagnosis and Treatment of Chronic Conditions* (pp. 301–314). Hershey, PA: IGI Global. doi:10.4018/978-1-4666-8828-5.ch014

Gopalan, V., Chan, E., & Ho, D. T. (2018). Deliberate Self-Harm and Suicide Ideology in Medical Students. In C. Smith (Ed.), *Exploring the Pressures of Medical Education From a Mental Health and Wellness Perspective* (pp. 122–143). Hershey, PA: IGI Global. doi:10.4018/978-1-5225-2811-1.ch005

Goswami, S., Dey, U., Roy, P., Ashour, A., & Dey, N. (2017). Medical Video Processing: Concept and Applications. In N. Dey, A. Ashour, & P. Patra (Eds.), *Feature Detectors and Motion Detection in Video Processing* (pp. 1–17). Hershey, PA: IGI Global. doi:10.4018/978-1-5225-1025-3.ch001

Goswami, S., Mahanta, K., Goswami, S., Jigdung, T., & Devi, T. P. (2018). Ageing and Cancer: The Epigenetic Basis, Alternative Treatment, and Care. In B. Prasad & S. Akbar (Eds.), *Handbook of Research on Geriatric Health, Treatment, and Care* (pp. 206–235). Hershey, PA: IGI Global. doi:10.4018/978-1-5225-3480-8.ch012

Gouva, M. I. (2017). The Psychological Impact of Medical Error on Patients, Family Members, and Health Professionals. In M. Riga (Ed.), *Impact of Medical Errors and Malpractice on Health Economics, Quality, and Patient Safety* (pp. 171–196). Hershey, PA: IGI Global. doi:10.4018/978-1-5225-2337-6.ch007

Gürcü, M., & Tengilimoğlu, D. (2017). Health Tourism-Based Destination Marketing. In A. Bayraktar & C. Uslay (Eds.), *Strategic Place Branding Methodologies and Theory for Tourist Attraction* (pp. 308–331). Hershey, PA: IGI Global. doi:10.4018/978-1-5225-0579-2.ch015

Gürsel, G. (2016). Mobility in Healthcare: M-Health. In A. Panagopoulos (Ed.), *Handbook of Research on Next Generation Mobile Communication Systems* (pp. 485–511). Hershey, PA: IGI Global. doi:10.4018/978-1-4666-8732-5.ch019

Gürsel, G. (2017). For Better Healthcare Mining Health Data. In S. Bhattacharyya, S. De, I. Pan, & P. Dutta (Eds.), *Intelligent Multidimensional Data Clustering and Analysis* (pp. 135–158). Hershey, PA: IGI Global. doi:10.4018/978-1-5225-1776-4.ch006

Guzman-Lugo, G., Lopez-Martinez, M., Lopez-Ramirez, B., & Avila-Vazquez, D. (2016). City-Driven Approach to Determine Health Services Based on Current User Location and Collaborative Information. *International Journal of Knowledge Society Research*, *7*(1), 53–62. doi:10.4018/IJKSR.2016010104

Gyaase, P. O., Darko-Lartey, R., William, H., & Borkloe, F. (2017). Towards an Integrated Electronic Medical Records System for Quality Healthcare in Ghana: An Exploratory Factor Analysis. *International Journal of Computers in Clinical Practice*, *2*(2), 38–55. doi:10.4018/IJCCP.2017070103

Handayani, P. W., Sandhyaduhita, P. I., Hidayanto, A. N., Pinem, A. A., Fajrina, H. R., Junus, K. M., ... Ayuningtyas, D. (2016). Integrated Hospital Information System Architecture Design in Indonesia. In T. Iyamu & A. Tatnall (Eds.), *Maximizing Healthcare Delivery and Management through Technology Integration* (pp. 207–236). Hershey, PA: IGI Global. doi:10.4018/978-1-4666-9446-0.ch013

Hanel, P. (2017). Is China Catching Up?: Health-Related Applications of Biotechnology. In T. Bas & J. Zhao (Eds.), *Comparative Approaches to Biotechnology Development and Use in Developed and Emerging Nations* (pp. 465–520). Hershey, PA: IGI Global. doi:10.4018/978-1-5225-1040-6.ch016

Harmsen, C. A., & Royle, R. N. (2016). St. Stephen's Hospital Hervey Bay: Study of Developing a Digital Hospital. In J. Moon & M. Galea (Eds.), *Improving Health Management through Clinical Decision Support Systems* (pp. 127–153). Hershey, PA: IGI Global. doi:10.4018/978-1-4666-9432-3.ch006

Harp, D., Shim, R. S., Johnson, J., Harp, J. A., Wilcox, W. C., & Wilcox, J. K. (2016). Race and Gender Inequalities in Medicine and Biomedical Research. In U. Thomas & J. Drake (Eds.), *Critical Research on Sexism and Racism in STEM Fields* (pp. 115–134). Hershey, PA: IGI Global. doi:10.4018/978-1-5225-0174-9.ch006

Hartman, A., & Brown, S. (2017). Synergism through Therapeutic Visual Arts. In V. Bryan & J. Bird (Eds.), *Healthcare Community Synergism between Patients, Practitioners, and Researchers* (pp. 29–48). Hershey, PA: IGI Global. doi:10.4018/978-1-5225-0640-9.ch002

Hazra, J., Chowdhury, A. R., & Dutta, P. (2016). Cluster Based Medical Image Registration Using Optimized Neural Network. In S. Bhattacharyya, P. Banerjee, D. Majumdar, & P. Dutta (Eds.), *Handbook of Research on Advanced Hybrid Intelligent Techniques and Applications* (pp. 551–581). Hershey, PA: IGI Global. doi:10.4018/978-1-4666-9474-3.ch018

Hung, S., Huang, W., Yen, D. C., Chang, S., & Lu, C. (2016). Effect of Information Service Competence and Contextual Factors on the Effectiveness of Strategic Information Systems Planning in Hospitals. *Journal of Global Information Management, 24*(1), 14–36. doi:10.4018/JGIM.2016010102

Iltchev, P., Śliwczyński, A., Szynkiewicz, P., & Marczak, M. (2016). Mobile Health Applications Assisting Patients with Chronic Diseases: Examples from Asthma Care. In A. Moumtzoglou (Ed.), *M-Health Innovations for Patient-Centered Care* (pp. 170–196). Hershey, PA: IGI Global. doi:10.4018/978-1-4666-9861-1.ch009

Iqbal, S., Ahmad, S., & Willis, I. (2017). Influencing Factors for Adopting Technology Enhanced Learning in the Medical Schools of Punjab, Pakistan. *International Journal of Information and Communication Technology Education*, *13*(3), 27–39. doi:10.4018/IJICTE.2017070103

Jagan, J., Dalkiliç, Y., & Samui, P. (2016). Utilization of SVM, LSSVM and GP for Predicting the Medical Waste Generation. In G. Hua (Ed.), *Smart Cities as a Solution for Reducing Urban Waste and Pollution* (pp. 224–251). Hershey, PA: IGI Global. doi:10.4018/978-1-5225-0302-6.ch008

Jagiello, K., Sosnowska, A., Mikolajczyk, A., & Puzyn, T. (2017). Nanomaterials in Medical Devices: Regulations' Review and Future Perspectives. *Journal of Nanotoxicology and Nanomedicine*, *2*(2), 1–11. doi:10.4018/JNN.2017070101

Jena, T. K. (2018). Skill Training Process in Medicine Through Distance Mode. In U. Pandey & V. Indrakanti (Eds.), *Optimizing Open and Distance Learning in Higher Education Institutions* (pp. 228–243). Hershey, PA: IGI Global. doi:10.4018/978-1-5225-2624-7.ch010

Jesus, Â., & Gomes, M. J. (2016). Web 2.0 Tools in Biomedical and Pharmaceutical Education: Updated Review and Commentary. In T. Gasmelseid (Ed.), *Advancing Pharmaceutical Processes and Tools for Improved Health Outcomes* (pp. 52–78). Hershey, PA: IGI Global. doi:10.4018/978-1-5225-0248-7.ch003

Joseph, V., & Miller, J. M. (2018). Medical Students' Perceived Stigma in Seeking Care: A Cultural Perspective. In C. Smith (Ed.), *Exploring the Pressures of Medical Education From a Mental Health and Wellness Perspective* (pp. 44–67). Hershey, PA: IGI Global. doi:10.4018/978-1-5225-2811-1.ch002

Joyce, B. L., & Swanberg, S. M. (2017). Using Backward Design for Competency-Based Undergraduate Medical Education. In J. Stefaniak (Ed.), *Advancing Medical Education Through Strategic Instructional Design* (pp. 53–76). Hershey, PA: IGI Global. doi:10.4018/978-1-5225-2098-6.ch003

Kaljo, K., & Jacques, L. (2018). Flipping the Medical School Classroom. In M. Khosrow-Pour, D.B.A. (Ed.), Encyclopedia of Information Science and Technology, Fourth Edition (pp. 5800-5809). Hershey, PA: IGI Global. doi:10.4018/978-1-5225-2255-3.ch504

Karon, R. (2016). Utilisation of Health Information Systems for Service Delivery in the Namibian Environment. In T. Iyamu & A. Tatnall (Eds.), *Maximizing Healthcare Delivery and Management through Technology Integration* (pp. 169–183). Hershey, PA: IGI Global. doi:10.4018/978-1-4666-9446-0.ch011

Kasina, H., Bahubalendruni, M. V., & Botcha, R. (2017). Robots in Medicine: Past, Present and Future. *International Journal of Manufacturing, Materials, and Mechanical Engineering*, *7*(4), 44–64. doi:10.4018/IJMMME.2017100104

Katehakis, D. G. (2018). Electronic Medical Record Implementation Challenges for the National Health System in Greece. *International Journal of Reliable and Quality E-Healthcare*, *7*(1), 16–30. doi:10.4018/IJRQEH.2018010102

Katz, A., & Shtub, A. (2016). Design and Build a Wizard of Oz (WOZ) Telemedicine Simulator Platform. In M. Cruz-Cunha, I. Miranda, R. Martinho, & R. Rijo (Eds.), *Encyclopedia of E-Health and Telemedicine* (pp. 128–141). Hershey, PA: IGI Global. doi:10.4018/978-1-4666-9978-6.ch011

Kaur, P. D., & Sharma, P. (2017). Success Dimensions of ICTs in Healthcare. In B. Singh (Ed.), *Computational Tools and Techniques for Biomedical Signal Processing* (pp. 149–173). Hershey, PA: IGI Global. doi:10.4018/978-1-5225-0660-7.ch007

Kaushik, P. (2018). Comorbidity of Medical and Psychiatric Disorders in Geriatric Population: Treatment and Care. In B. Prasad & S. Akbar (Eds.), *Handbook of Research on Geriatric Health, Treatment, and Care* (pp. 448–474). Hershey, PA: IGI Global. doi:10.4018/978-1-5225-3480-8.ch025

Khachane, M. Y. (2017). Organ-Based Medical Image Classification Using Support Vector Machine. *International Journal of Synthetic Emotions*, *8*(1), 18–30. doi:10.4018/IJSE.2017010102

Kirci, P. (2018). Intelligent Techniques for Analysis of Big Data About Healthcare and Medical Records. In N. Shah & M. Mittal (Eds.), *Handbook of Research on Promoting Business Process Improvement Through Inventory Control Techniques* (pp. 559–582). Hershey, PA: IGI Global. doi:10.4018/978-1-5225-3232-3.ch029

Kldiashvili, E. (2016). Cloud Computing as the Useful Resource for Application of the Medical Information System for Quality Assurance Purposes. *International Journal of Computers in Clinical Practice*, *1*(2), 1–23. doi:10.4018/IJCCP.2016070101

Kldiashvili, E. (2018). Cloud Approach for the Medical Information System: MIS on Cloud. In M. Khosrow-Pour (Ed.), *Incorporating Nature-Inspired Paradigms in Computational Applications* (pp. 238–261). Hershey, PA: IGI Global. doi:10.4018/978-1-5225-5020-4.ch008

Kldiashvili, E., Burduli, A., Ghortlishvili, G., & Sheklashvili, I. (2016). Georgian Experience in Telecytology. In M. Cruz-Cunha, I. Miranda, R. Martinho, & R. Rijo (Eds.), *Encyclopedia of E-Health and Telemedicine* (pp. 62–71). Hershey, PA: IGI Global. doi:10.4018/978-1-4666-9978-6.ch006

Ko, H., Mesicek, L., Choi, J., Choi, J., & Hwang, S. (2018). A Study on Secure Contents Strategies for Applications With DRM on Cloud Computing. *International Journal of Cloud Applications and Computing*, *8*(1), 143–153. doi:10.4018/IJCAC.2018010107

Komendziński, T., Dreszer-Drogorób, J., Mikołajewska, E., Mikołajewski, D., & Bałaj, B. (2016). Interdisciplinary Education for Research and Everyday Clinical Practice: Lessons Learned from InteRDoCTor Project. In A. Rosiek & K. Leksowski (Eds.), *Organizational Culture and Ethics in Modern Medicine* (pp. 78–110). Hershey, PA: IGI Global. doi:10.4018/978-1-4666-9658-7.ch004

Komendziński, T., Mikołajewska, E., & Mikołajewski, D. (2018). Cross-Cultural Decision-Making in Healthcare: Theory and Practical Application in Real Clinical Conditions. In A. Rosiek-Kryszewska & K. Leksowski (Eds.), *Healthcare Administration for Patient Safety and Engagement* (pp. 276–298). Hershey, PA: IGI Global. doi:10.4018/978-1-5225-3946-9.ch015

Konecny, L. T. (2018). Medical School Wellness Initiatives. In C. Smith (Ed.), *Exploring the Pressures of Medical Education From a Mental Health and Wellness Perspective* (pp. 209–228). Hershey, PA: IGI Global. doi:10.4018/978-1-5225-2811-1.ch009

Konieczna, A., & Słomkowski, P. (2016). Confrontation of Human Rights in Daily Clinical Situations. In A. Rosiek & K. Leksowski (Eds.), *Organizational Culture and Ethics in Modern Medicine* (pp. 255–281). Hershey, PA: IGI Global. doi:10.4018/978-1-4666-9658-7.ch011

Kromrei, H., Solomonson, W. L., & Juzych, M. S. (2017). Teaching Residents How to Teach. In J. Stefaniak (Ed.), *Advancing Medical Education Through Strategic Instructional Design* (pp. 164–185). Hershey, PA: IGI Global. doi:10.4018/978-1-5225-2098-6.ch008

Kulkarni, S., Savyanavar, A., Kulkarni, P., Stranieri, A., & Ghorpade, V. (2018). Framework for Integration of Medical Image and Text-Based Report Retrieval to Support Radiological Diagnosis. In M. Kolekar & V. Kumar (Eds.), *Biomedical Signal and Image Processing in Patient Care* (pp. 86–122). Hershey, PA: IGI Global. doi:10.4018/978-1-5225-2829-6.ch006

Kumar, A., & Sarkar, B. K. (2018). Performance Analysis of Nature-Inspired Algorithms-Based Bayesian Prediction Models for Medical Data Sets. In U. Singh, A. Tiwari, & R. Singh (Eds.), *Soft-Computing-Based Nonlinear Control Systems Design* (pp. 134–155). Hershey, PA: IGI Global. doi:10.4018/978-1-5225-3531-7.ch007

Kurtz, R. S. (2016). Fortitude: A Study of African Americans in Surgery in New York City. In U. Thomas & J. Drake (Eds.), *Critical Research on Sexism and Racism in STEM Fields* (pp. 153–169). Hershey, PA: IGI Global. doi:10.4018/978-1-5225-0174-9.ch009

Labbadi, W., & Akaichi, J. (2017). Efficient Algorithm for Answering Fuzzy Medical Requests in Pervasive Healthcare Information Systems. *International Journal of Healthcare Information Systems and Informatics*, *12*(2), 46–64. doi:10.4018/IJHISI.2017040103

Lagumdzija, A., & Swing, V. K. (2017). Health, Digitalization, and Individual Empowerment. In F. Topor (Ed.), *Handbook of Research on Individualism and Identity in the Globalized Digital Age* (pp. 380–402). Hershey, PA: IGI Global. doi:10.4018/978-1-5225-0522-8.ch017

Lamey, T. W., & Davidson-Shivers, G. V. (2017). Instructional Strategies and Sequencing. In J. Stefaniak (Ed.), *Advancing Medical Education Through Strategic Instructional Design* (pp. 30–52). Hershey, PA: IGI Global. doi:10.4018/978-1-5225-2098-6.ch002

Leon, G. (2017). The Role of Forensic Medicine in Medical Errors. In M. Riga (Ed.), *Impact of Medical Errors and Malpractice on Health Economics, Quality, and Patient Safety* (pp. 144–170). Hershey, PA: IGI Global. doi:10.4018/978-1-5225-2337-6.ch006

Lidia, B., Federica, G., & Maria, V. E. (2016). The Patient-Centered Medicine as the Theoretical Framework for Patient Engagement. In G. Graffigna (Ed.), *Promoting Patient Engagement and Participation for Effective Healthcare Reform* (pp. 25–39). Hershey, PA: IGI Global. doi:10.4018/978-1-4666-9992-2.ch002

Love, L., & McDowelle, D. (2018). Developing a Comprehensive Wellness Program for Medical Students. In C. Smith (Ed.), *Exploring the Pressures of Medical Education From a Mental Health and Wellness Perspective* (pp. 190–208). Hershey, PA: IGI Global. doi:10.4018/978-1-5225-2811-1.ch008

Lovell, K. L. (2017). Development and Evaluation of Neuroscience Computer-Based Modules for Medical Students: Instructional Design Principles and Effectiveness. In J. Stefaniak (Ed.), *Advancing Medical Education Through Strategic Instructional Design* (pp. 262–276). Hershey, PA: IGI Global. doi:10.4018/978-1-5225-2098-6.ch013

Lubin, R., & Hamlin, M. D. (2018). Medical Student Burnout: A Social Cognitive Learning Perspective on Medical Student Mental Health and Wellness. In C. Smith (Ed.), *Exploring the Pressures of Medical Education From a Mental Health and Wellness Perspective* (pp. 92–121). Hershey, PA: IGI Global. doi:10.4018/978-1-5225-2811-1.ch004

Luk, C. Y. (2018). Moving Towards Universal Health Coverage: Challenges for the Present and Future in China. In B. Fong, A. Ng, & P. Yuen (Eds.), *Sustainable Health and Long-Term Care Solutions for an Aging Population* (pp. 19–45). Hershey, PA: IGI Global. doi:10.4018/978-1-5225-2633-9.ch002

Mahat, M., & Pettigrew, A. (2017). The Regulatory Environment of Non-Profit Higher Education and Research Institutions and Its Implications for Managerial Strategy: The Case of Medical Education and Research. In L. West & A. Worthington (Eds.), *Handbook of Research on Emerging Business Models and Managerial Strategies in the Nonprofit Sector* (pp. 336–351). Hershey, PA: IGI Global. doi:10.4018/978-1-5225-2537-0.ch017

Maitra, I. K., & Bandhyopadhyaay, S. K. (2017). Adaptive Edge Detection Method towards Features Extraction from Diverse Medical Imaging Technologies. In S. Bhattacharyya, S. De, I. Pan, & P. Dutta (Eds.), *Intelligent Multidimensional Data Clustering and Analysis* (pp. 159–192). Hershey, PA: IGI Global. doi:10.4018/978-1-5225-1776-4.ch007

Mangu, V. P. (2017). Mobile Health Care: A Technology View. In C. Bhatt & S. Peddoju (Eds.), *Cloud Computing Systems and Applications in Healthcare* (pp. 1–18). Hershey, PA: IGI Global. doi:10.4018/978-1-5225-1002-4.ch001

Manirabona, A., Fourati, L. C., & Boudjit, S. (2017). Investigation on Healthcare Monitoring Systems: Innovative Services and Applications. *International Journal of E-Health and Medical Communications*, *8*(1), 1–18. doi:10.4018/IJEHMC.2017010101

Mannai, M. M., & Karâa, W. B. (2016). Biomedical Image Processing Overview. In W. Karâa & N. Dey (Eds.), *Biomedical Image Analysis and Mining Techniques for Improved Health Outcomes* (pp. 1–12). Hershey, PA: IGI Global. doi:10.4018/978-1-4666-8811-7.ch001

Manzoor, A. (2016). RFID in Health Care-Building Smart Hospitals for Quality Healthcare. *International Journal of User-Driven Healthcare*, *6*(2), 21–45. doi:10.4018/IJUDH.2016070102

Marwan, M., Kartit, A., & Ouahmane, H. (2018). A Framework to Secure Medical Image Storage in Cloud Computing Environment. *Journal of Electronic Commerce in Organizations*, *16*(1), 1–16. doi:10.4018/JECO.2018010101

Masoud, M. P., Nejad, M. K., Darebaghi, H., Chavoshi, M., & Farahani, M. (2018). The Decision Support System and Conventional Method of Telephone Triage by Nurses in Emergency Medical Services: A Comparative Investigation. *International Journal of E-Business Research*, *14*(1), 77–88. doi:10.4018/IJEBR.2018010105

McDonald, W. G., Martin, M., & Salzberg, L. D. (2018). From Medical Student to Medical Resident: Graduate Medical Education and Mental Health in the United States. In C. Smith (Ed.), *Exploring the Pressures of Medical Education From a Mental Health and Wellness Perspective* (pp. 145–169). Hershey, PA: IGI Global. doi:10.4018/978-1-5225-2811-1.ch006

Medhekar, A. (2017). The Role of Social Media for Knowledge Dissemination in Medical Tourism: A Case of India. In R. Chugh (Ed.), *Harnessing Social Media as a Knowledge Management Tool* (pp. 25–54). Hershey, PA: IGI Global. doi:10.4018/978-1-5225-0495-5.ch002

Medhekar, A., & Haq, F. (2018). Urbanization and New Jobs Creation in Healthcare Services in India: Challenges and Opportunities. In U. Benna & I. Benna (Eds.), *Urbanization and Its Impact on Socio-Economic Growth in Developing Regions* (pp. 198–218). Hershey, PA: IGI Global. doi:10.4018/978-1-5225-2659-9.ch010

Mehta, G., Dutta, M. K., & Kim, P. S. (2016). An Efficient and Lossless Cryptosystem for Security in Tele-Ophthalmology Applications Using Chaotic Theory. *International Journal of E-Health and Medical Communications*, *7*(4), 28–47. doi:10.4018/IJEHMC.2016100102

Mehta, P. (2017). Framework of Indian Healthcare System and its Challenges: An Insight. In V. Bryan & J. Bird (Eds.), *Healthcare Community Synergism between Patients, Practitioners, and Researchers* (pp. 247–271). Hershey, PA: IGI Global. doi:10.4018/978-1-5225-0640-9.ch011

Metelmann, B., & Metelmann, C. (2016). M-Health in Prehospital Emergency Medicine: Experiences from the EU funded Project LiveCity. In A. Moumtzoglou (Ed.), *M-Health Innovations for Patient-Centered Care* (pp. 197–212). Hershey, PA: IGI Global. doi:10.4018/978-1-4666-9861-1.ch010

Mi, M. (2017). Informal Learning in Medical Education. In J. Stefaniak (Ed.), *Advancing Medical Education Through Strategic Instructional Design* (pp. 225–244). Hershey, PA: IGI Global. doi:10.4018/978-1-5225-2098-6.ch011

Miglioretti, M., Mariani, F., & Vecchio, L. (2016). Could Patient Engagement Promote a Health System Free From Malpractice Litigation Risk? In G. Graffigna (Ed.), *Promoting Patient Engagement and Participation for Effective Healthcare Reform* (pp. 240–264). Hershey, PA: IGI Global. doi:10.4018/978-1-4666-9992-2.ch012

Misra, S. C., & Bisui, S. (2016). Feasibility of Large Scale Implementation of Personalized Medicine in the Current Scenario. *International Journal of E-Health and Medical Communications*, *7*(2), 30–49. doi:10.4018/IJEHMC.2016040103

Misra, S. C., Bisui, S., & Fantazy, K. (2016). Identifying Critical Changes in Adoption of Personalized Medicine (PM) in Healthcare Management. *International Journal of E-Health and Medical Communications*, *7*(3), 1–15. doi:10.4018/IJEHMC.2016070101

Mohammadian, M., Hatzinakos, D., Spachos, P., & Jentzsh, R. (2016). An Intelligent and Secure Framework for Wireless Information Technology in Healthcare for User and Data Classification in Hospitals. In V. Wang (Ed.), *Handbook of Research on Advancing Health Education through Technology* (pp. 452–479). Hershey, PA: IGI Global. doi:10.4018/978-1-4666-9494-1.ch019

Mokeddem, S., & Atmani, B. (2016). Assessment of Clinical Decision Support Systems for Predicting Coronary Heart Disease. *International Journal of Operations Research and Information Systems*, *7*(3), 57–73. doi:10.4018/IJORIS.2016070104

Mokrzycka, A., & Kowalska-Bobko, I. (2016). Public Health Legislation and Patient's Rights: Health2020 Strategy, European Perspective. In A. Rosiek & K. Leksowski (Eds.), *Organizational Culture and Ethics in Modern Medicine* (pp. 298–322). Hershey, PA: IGI Global. doi:10.4018/978-1-4666-9658-7.ch013

Moon, J. D., & Galea, M. P. (2016). Overview of Clinical Decision Support Systems in Healthcare. In J. Moon & M. Galea (Eds.), *Improving Health Management through Clinical Decision Support Systems* (pp. 1–27). Hershey, PA: IGI Global. doi:10.4018/978-1-4666-9432-3.ch001

Mosadeghrad, A. M., & Woldemichael, A. (2017). Application of Quality Management in Promoting Patient Safety and Preventing Medical Errors. In M. Riga (Ed.), *Impact of Medical Errors and Malpractice on Health Economics, Quality, and Patient Safety* (pp. 91–112). Hershey, PA: IGI Global. doi:10.4018/978-1-5225-2337-6.ch004

Moumtzoglou, A. (2017). Digital Medicine: The Quality Standpoint. In A. Moumtzoglou (Ed.), *Design, Development, and Integration of Reliable Electronic Healthcare Platforms* (pp. 179–195). Hershey, PA: IGI Global. doi:10.4018/978-1-5225-1724-5.ch011

Moumtzoglou, A., & Pouliakis, A. (2018). Population Health Management and the Science of Individuality. *International Journal of Reliable and Quality E-Healthcare, 7*(2), 1–26. doi:10.4018/IJRQEH.2018040101

Mpofu, C. (2016). International Medical Experiences Outbound New Zealand: An Economic and Medical Workforce Strategy. In D. Velliaris & D. Coleman-George (Eds.), *Handbook of Research on Study Abroad Programs and Outbound Mobility* (pp. 446–469). Hershey, PA: IGI Global. doi:10.4018/978-1-5225-0169-5.ch018

Mukhtar, W. F., & Abuelyaman, E. S. (2017). Opportunities and Challenges of Big Data in Healthcare. In N. Wickramasinghe (Ed.), *Handbook of Research on Healthcare Administration and Management* (pp. 47–58). Hershey, PA: IGI Global. doi:10.4018/978-1-5225-0920-2.ch004

Munugala, S., Brar, G. K., Syed, A., Mohammad, A., & Halgamuge, M. N. (2018). The Much Needed Security and Data Reforms of Cloud Computing in Medical Data Storage. In M. Lytras & P. Papadopoulou (Eds.), *Applying Big Data Analytics in Bioinformatics and Medicine* (pp. 99–113). Hershey, PA: IGI Global. doi:10.4018/978-1-5225-2607-0.ch005

Narasimhamurthy, A. (2016). An Overview of Machine Learning in Medical Image Analysis: Trends in Health Informatics. In N. Dey & A. Ashour (Eds.), *Classification and Clustering in Biomedical Signal Processing* (pp. 23–45). Hershey, PA: IGI Global. doi:10.4018/978-1-5225-0140-4.ch002

Ngara, R. (2017). Multiple Voices, Multiple Paths: Towards Dialogue between Western and Indigenous Medical Knowledge Systems. In P. Ngulube (Ed.), *Handbook of Research on Theoretical Perspectives on Indigenous Knowledge Systems in Developing Countries* (pp. 332–358). Hershey, PA: IGI Global. doi:10.4018/978-1-5225-0833-5.ch015

O'Connor, Y., & Heavin, C. (2018). Defining and Characterising the Landscape of eHealth. In M. Khosrow-Pour, D.B.A. (Ed.), Encyclopedia of Information Science and Technology, Fourth Edition (pp. 5864-5875). Hershey, PA: IGI Global. doi:10.4018/978-1-5225-2255-3.ch510

Ochonogor, W. C., & Okite-Amughoro, F. A. (2018). Building an Effective Digital Library in a University Teaching Hospital (UTH) in Nigeria. In A. Tella & T. Kwanya (Eds.), *Handbook of Research on Managing Intellectual Property in Digital Libraries* (pp. 184–204). Hershey, PA: IGI Global. doi:10.4018/978-1-5225-3093-0.ch010

Olaniran, B. A. (2016). ICTs, E-health, and Multidisciplinary Healthcare Teams: Promises and Challenges. *International Journal of Privacy and Health Information Management*, *4*(2), 62–75. doi:10.4018/IJPHIM.2016070105

Olaniran, B. A. (2016). ICT Use and Multidisciplinary Healthcare Teams in the Age of e-Health. *International Journal of Reliable and Quality E-Healthcare*, *5*(1), 18–31. doi:10.4018/IJRQEH.2016010102

Olivares, S. L., Cruz, A. G., Cabrera, M. V., Regalado, A. I., & García, J. E. (2017). An Assessment Study of Quality Model for Medical Schools in Mexico. In S. Mukerji & P. Tripathi (Eds.), *Handbook of Research on Administration, Policy, and Leadership in Higher Education* (pp. 404–439). Hershey, PA: IGI Global. doi:10.4018/978-1-5225-0672-0.ch016

Omidoyin, E. O., Opeke, R. O., & Osagbemi, G. K. (2016). Utilization Pattern and Privacy Issues in the use of Health Records for Research Practice by Doctors: Selected Nigerian Teaching Hospitals as Case Study. *International Journal of Privacy and Health Information Management*, *4*(1), 1–11. doi:10.4018/IJPHIM.2016010101

Omoruyi, E. A., & Omidele, F. (2018). Resident Physician and Medical Academic Faculty Burnout: A Review of Current Literature. In C. Smith (Ed.), *Exploring the Pressures of Medical Education From a Mental Health and Wellness Perspective* (pp. 171–189). Hershey, PA: IGI Global. doi:10.4018/978-1-5225-2811-1.ch007

Osop, H., & Sahama, T. (2016). Data-Driven and Practice-Based Evidence: Design and Development of Efficient and Effective Clinical Decision Support System. In J. Moon & M. Galea (Eds.), *Improving Health Management through Clinical Decision Support Systems* (pp. 295–328). Hershey, PA: IGI Global. doi:10.4018/978-1-4666-9432-3.ch014

Ossowski, R., & Izdebski, P. (2016). Ethical Aspects of Talking to a Patient. In A. Rosiek & K. Leksowski (Eds.), *Organizational Culture and Ethics in Modern Medicine* (pp. 203–235). Hershey, PA: IGI Global. doi:10.4018/978-1-4666-9658-7.ch009

Pandian, P. S. (2016). An Overview of Telemedicine Technologies for Healthcare Applications. *International Journal of Biomedical and Clinical Engineering, 5*(2), 29–52. doi:10.4018/IJBCE.2016070103

Papadopoulou, P., Lytras, M., & Marouli, C. (2018). Bioinformatics as Applied to Medicine: Challenges Faced Moving from Big Data to Smart Data to Wise Data. In M. Lytras & P. Papadopoulou (Eds.), *Applying Big Data Analytics in Bioinformatics and Medicine* (pp. 1–25). Hershey, PA: IGI Global. doi:10.4018/978-1-5225-2607-0.ch001

Paraskou, A., & George, B. P. (2018). An Overview of Reproductive Tourism. In *Legal and Economic Considerations Surrounding Reproductive Tourism: Emerging Research and Opportunities* (pp. 1–17). Hershey, PA: IGI Global. doi:10.4018/978-1-5225-2694-0.ch001

Park, S., & Moon, J. (2016). Strategic Approach towards Clinical Information Security. In J. Moon & M. Galea (Eds.), *Improving Health Management through Clinical Decision Support Systems* (pp. 329–359). Hershey, PA: IGI Global. doi:10.4018/978-1-4666-9432-3.ch015

Pereira, D., Castro, A., Gomes, P., Areias, J. C., Reis, Z. S., Coimbra, M. T., & Cruz-Correia, R. (2016). Digital Auscultation: Challenges and Perspectives. In M. Cruz-Cunha, I. Miranda, R. Martinho, & R. Rijo (Eds.), *Encyclopedia of E-Health and Telemedicine* (pp. 910–927). Hershey, PA: IGI Global. doi:10.4018/978-1-4666-9978-6.ch070

Peters, R. A., & Cruz, M. (2016). The States as Generators of Incremental Change in American Health Care Policy: 1935 to 1965. In R. Gholipour & K. Rouzbehani (Eds.), *Social, Economic, and Political Perspectives on Public Health Policy-Making* (pp. 86–114). Hershey, PA: IGI Global. doi:10.4018/978-1-4666-9944-1.ch005

Phuritsabam, B., & Devi, A. B. (2017). Information Seeking Behavior of Medical Scientists at Jawaharlal Nehru Institute of Medical Science: A Study. In S. Ram (Ed.), *Library and Information Services for Bioinformatics Education and Research* (pp. 177–187). Hershey, PA: IGI Global. doi:10.4018/978-1-5225-1871-6.ch010

Pieczka, B. (2018). Management of Risk and Adverse Events in Medical Entities. In A. Rosiek-Kryszewska & K. Leksowski (Eds.), *Healthcare Administration for Patient Safety and Engagement* (pp. 31–46). Hershey, PA: IGI Global. doi:10.4018/978-1-5225-3946-9.ch003

Poduval, J. (2016). Curriculum Development. In *Optimizing Medicine Residency Training Programs* (pp. 45–102). Hershey, PA: IGI Global. doi:10.4018/978-1-4666-9527-6.ch003

Poduval, J. (2016). Ethics and Professionalism. In *Optimizing Medicine Residency Training Programs* (pp. 103–133). Hershey, PA: IGI Global. doi:10.4018/978-1-4666-9527-6.ch004

Poduval, J. (2016). Management Skills and Leadership. In *Optimizing Medicine Residency Training Programs* (pp. 182–205). Hershey, PA: IGI Global. doi:10.4018/978-1-4666-9527-6.ch008

Poduval, J. (2016). Medicine Residency Training. In *Optimizing Medicine Residency Training Programs* (pp. 1–27). Hershey, PA: IGI Global. doi:10.4018/978-1-4666-9527-6.ch001

Poduval, J. (2016). Personal Issues. In *Optimizing Medicine Residency Training Programs* (pp. 206–221). Hershey, PA: IGI Global. doi:10.4018/978-1-4666-9527-6.ch009

Poduval, J. (2017). Medical Errors: Impact on Health Care Quality. In M. Riga (Ed.), *Impact of Medical Errors and Malpractice on Health Economics, Quality, and Patient Safety* (pp. 33–60). Hershey, PA: IGI Global. doi:10.4018/978-1-5225-2337-6.ch002

Politis, D., Stagiopoulos, P., Aidona, S., Kyriafinis, G., & Constantinidis, I. (2018). Autonomous Learning and Skill Accreditation: A Paradigm for Medical Studies. In A. Kumar (Ed.), *Optimizing Student Engagement in Online Learning Environments* (pp. 266–296). Hershey, PA: IGI Global. doi:10.4018/978-1-5225-3634-5.ch012

Pomares-Quimbaya, A., Gonzalez, R. A., Quintero, S., Muñoz, O. M., Bohórquez, W. R., García, O. M., & Londoño, D. (2016). A Review of Existing Applications and Techniques for Narrative Text Analysis in Electronic Medical Records. In M. Cruz-Cunha, I. Miranda, R. Martinho, & R. Rijo (Eds.), *Encyclopedia of E-Health and Telemedicine* (pp. 796–811). Hershey, PA: IGI Global. doi:10.4018/978-1-4666-9978-6.ch062

Pomares-Quimbaya, A., González, R. A., Sierra, A., Daza, J. C., Muñoz, O., García, A., ... Bohórquez, W. R. (2017). ICT for Enabling the Quality Evaluation of Health Care Services: A Case Study in a General Hospital. In A. Moumtzoglou (Ed.), *Design, Development, and Integration of Reliable Electronic Healthcare Platforms* (pp. 196–210). Hershey, PA: IGI Global. doi:10.4018/978-1-5225-1724-5.ch012

Pouliakis, A., Margari, N., Karakitsou, E., Archondakis, S., & Karakitsos, P. (2018). Emerging Technologies Serving Cytopathology: Big Data, the Cloud, and Mobile Computing. In I. El Naqa (Ed.), *Emerging Developments and Practices in Oncology* (pp. 114–152). Hershey, PA: IGI Global. doi:10.4018/978-1-5225-3085-5.ch005

Queirós, A., Silva, A. G., Ferreira, A., Caravau, H., Cerqueira, M., & Rocha, N. P. (2016). Assessing Mobile Applications Considered Medical Devices. In M. Cruz-Cunha, I. Miranda, R. Martinho, & R. Rijo (Eds.), *Encyclopedia of E-Health and Telemedicine* (pp. 111–127). Hershey, PA: IGI Global. doi:10.4018/978-1-4666-9978-6.ch010

Raffaeli, L., Spinsante, S., & Gambi, E. (2016). Integrated Smart TV-Based Personal e-Health System. *International Journal of E-Health and Medical Communications*, *7*(1), 48–64. doi:10.4018/IJEHMC.2016010103

Rai, A., Kothari, R., & Singh, D. P. (2017). Assessment of Available Technologies for Hospital Waste Management: A Need for Society. In R. Singh, A. Singh, & V. Srivastava (Eds.), *Environmental Issues Surrounding Human Overpopulation* (pp. 172–188). Hershey, PA: IGI Global. doi:10.4018/978-1-5225-1683-5.ch010

Ramamoorthy, S., & Sivasubramaniam, R. (2018). Image Processing Including Medical Liver Imaging: Medical Image Processing from Big Data Perspective, Ultrasound Liver Images, Challenges. In M. Lytras & P. Papadopoulou (Eds.), *Applying Big Data Analytics in Bioinformatics and Medicine* (pp. 380–392). Hershey, PA: IGI Global. doi:10.4018/978-1-5225-2607-0.ch016

Rathor, G. P., & Gupta, S. K. (2017). Improving Multimodality Image Fusion through Integrate AFL and Wavelet Transform. In V. Tiwari, B. Tiwari, R. Thakur, & S. Gupta (Eds.), *Pattern and Data Analysis in Healthcare Settings* (pp. 143–157). Hershey, PA: IGI Global. doi:10.4018/978-1-5225-0536-5.ch008

Rawat, D. B., & Bhattacharya, S. (2016). Wireless Body Area Network for Healthcare Applications. In N. Meghanathan (Ed.), *Advanced Methods for Complex Network Analysis* (pp. 343–358). Hershey, PA: IGI Global. doi:10.4018/978-1-4666-9964-9.ch014

Rea, P. M. (2016). Advances in Anatomical and Medical Visualisation. In M. Pinheiro & D. Simões (Eds.), *Handbook of Research on Engaging Digital Natives in Higher Education Settings* (pp. 244–264). Hershey, PA: IGI Global. doi:10.4018/978-1-5225-0039-1.ch011

Rexhepi, H., & Persson, A. (2017). Challenges to Implementing IT Support for Evidence Based Practice Among Nurses and Assistant Nurses: A Qualitative Study. *Journal of Electronic Commerce in Organizations*, *15*(2), 61–76. doi:10.4018/JECO.2017040105

Reychav, I., & Azuri, J. (2016). Including Elderly Patients in Decision Making via Electronic Health Literacy. In M. Cruz-Cunha, I. Miranda, R. Martinho, & R. Rijo (Eds.), *Encyclopedia of E-Health and Telemedicine* (pp. 241–249). Hershey, PA: IGI Global. doi:10.4018/978-1-4666-9978-6.ch020

Richards, D., & Caldwell, P. H. (2016). Gamification to Improve Adherence to Clinical Treatment Advice: Improving Adherence to Clinical Treatment. In D. Novák, B. Tulu, & H. Brendryen (Eds.), *Handbook of Research on Holistic Perspectives in Gamification for Clinical Practice* (pp. 47–77). Hershey, PA: IGI Global. doi:10.4018/978-1-4666-9522-1.ch004

Rissman, B. (2016). Medical Conditions Associated with NLD. In B. Rissman (Ed.), *Medical and Educational Perspectives on Nonverbal Learning Disability in Children and Young Adults* (pp. 27–66). Hershey, PA: IGI Global. doi:10.4018/978-1-4666-9539-9.ch002

Rosiek, A. (2018). The Assessment of Actions of the Environment and the Impact of Preventive Medicine for Public Health in Poland. In A. Rosiek-Kryszewska & K. Leksowski (Eds.), *Healthcare Administration for Patient Safety and Engagement* (pp. 106–119). Hershey, PA: IGI Global. doi:10.4018/978-1-5225-3946-9.ch006

Rosiek, A., Leksowski, K., Goch, A., Rosiek-Kryszewska, A., & Leksowski, Ł. (2016). Medical Treatment and Difficult Ethical Decisions in Interdisciplinary Hospital Teams. In A. Rosiek & K. Leksowski (Eds.), *Organizational Culture and Ethics in Modern Medicine* (pp. 121–153). Hershey, PA: IGI Global. doi:10.4018/978-1-4666-9658-7.ch006

Rosiek, A., & Rosiek-Kryszewska, A. (2018). Managed Healthcare: Doctor Life Satisfaction and Its Impact on the Process of Communicating With the Patient. In A. Rosiek-Kryszewska & K. Leksowski (Eds.), *Healthcare Administration for Patient Safety and Engagement* (pp. 244–261). Hershey, PA: IGI Global. doi:10.4018/978-1-5225-3946-9.ch013

Rosiek-Kryszewska, A., Leksowski, Ł., Rosiek, A., Leksowski, K., & Goch, A. (2016). Clinical Communication in the Aspect of Development of New Technologies and E-Health in the Doctor-Patient Relationship. In A. Rosiek & K. Leksowski (Eds.), *Organizational Culture and Ethics in Modern Medicine* (pp. 18–51). Hershey, PA: IGI Global. doi:10.4018/978-1-4666-9658-7.ch002

Rosiek-Kryszewska, A., & Rosiek, A. (2018). The Involvement of the Patient and his Perspective Evaluation of the Quality of Healthcare. In A. Rosiek-Kryszewska & K. Leksowski (Eds.), *Healthcare Administration for Patient Safety and Engagement* (pp. 121–144). Hershey, PA: IGI Global. doi:10.4018/978-1-5225-3946-9.ch007

Rosiek-Kryszewska, A., & Rosiek, A. (2018). The Impact of Management and Leadership Roles in Building Competitive Healthcare Units. In A. Rosiek-Kryszewska & K. Leksowski (Eds.), *Healthcare Administration for Patient Safety and Engagement* (pp. 13–30). Hershey, PA: IGI Global. doi:10.4018/978-1-5225-3946-9.ch002

Roşu, S. M., Păvăloiu, I. B., Dragoi, G., Apostol, C. G., & Munteanu, D. (2016). Telemedicine Based on LMDS in the Urban/Metropolitan Area: A Romanian Case Study. In M. Cruz-Cunha, I. Miranda, R. Martinho, & R. Rijo (Eds.), *Encyclopedia of E-Health and Telemedicine* (pp. 96–109). Hershey, PA: IGI Global. doi:10.4018/978-1-4666-9978-6.ch009

Rouzbehani, K. (2017). Health Policy Implementation: Moving Beyond Its Barriers in United States. In N. Wickramasinghe (Ed.), *Handbook of Research on Healthcare Administration and Management* (pp. 541–552). Hershey, PA: IGI Global. doi:10.4018/978-1-5225-0920-2.ch032

Ruiz, I. M., Cohen, D. S., & Marco, Á. M. (2016). ISO/IEEE11073 Family of Standards: Trends and Applications on E-Health Monitoring. In M. Cruz-Cunha, I. Miranda, R. Martinho, & R. Rijo (Eds.), *Encyclopedia of E-Health and Telemedicine* (pp. 646–660). Hershey, PA: IGI Global. doi:10.4018/978-1-4666-9978-6.ch050

Sam, S. (2017). Mobile Phones and Expanding Human Capabilities in Plural Health Systems. In K. Moahi, K. Bwalya, & P. Sebina (Eds.), *Health Information Systems and the Advancement of Medical Practice in Developing Countries* (pp. 93–114). Hershey, PA: IGI Global. doi:10.4018/978-1-5225-2262-1.ch006

Sankaranarayanan, S., & Ganesan, S. (2016). Applications of Intelligent Agents in Health Sector-A Review. *International Journal of E-Health and Medical Communications*, *7*(1), 1–30. doi:10.4018/IJEHMC.2016010101

Santos-Trigo, M., Suaste, E., & Figuerola, P. (2018). Technology Design and Routes for Tool Appropriation in Medical Practices. In M. Khosrow-Pour, D.B.A. (Ed.), Encyclopedia of Information Science and Technology, Fourth Edition (pp. 3794-3804). Hershey, PA: IGI Global. doi:10.4018/978-1-5225-2255-3.ch329

Sarivougioukas, J., Vagelatos, A., Parsopoulos, K. E., & Lagaris, I. E. (2018). Home UbiHealth. In M. Khosrow-Pour, D.B.A. (Ed.), Encyclopedia of Information Science and Technology, Fourth Edition (pp. 7765-7774). Hershey, PA: IGI Global. doi:10.4018/978-1-5225-2255-3.ch675

Sarkar, B. K. (2017). Big Data and Healthcare Data: A Survey. *International Journal of Knowledge-Based Organizations*, *7*(4), 50–77. doi:10.4018/IJKBO.2017100104

Saxena, K., & Banodha, U. (2017). An Essence of the SOA on Healthcare. In R. Bhadoria, N. Chaudhari, G. Tomar, & S. Singh (Eds.), *Exploring Enterprise Service Bus in the Service-Oriented Architecture Paradigm* (pp. 283–304). Hershey, PA: IGI Global. doi:10.4018/978-1-5225-2157-0.ch018

Schmeida, M., & McNeal, R. (2016). Consulting Online Healthcare Information: E-Caregivers as Knowledgeable Decision Makers. *International Journal of Computers in Clinical Practice*, *1*(1), 42–52. doi:10.4018/IJCCP.2016010104

Sen, K., & Ghosh, K. (2018). Incorporating Global Medical Knowledge to Solve Healthcare Problems: A Framework for a Crowdsourcing System. *International Journal of Healthcare Information Systems and Informatics*, *13*(1), 1–14. doi:10.4018/IJHISI.2018010101

Shakdher, A., & Pandey, K. (2017). REDAlert+: Medical/Fire Emergency and Warning System using Android Devices. *International Journal of E-Health and Medical Communications*, *8*(1), 37–51. doi:10.4018/IJEHMC.2017010103

Shekarian, E., Abdul-Rashid, S. H., & Olugu, E. U. (2017). An Integrated Fuzzy VIKOR Method for Performance Management in Healthcare. In M. Tavana, K. Szabat, & K. Puranam (Eds.), *Organizational Productivity and Performance Measurements Using Predictive Modeling and Analytics* (pp. 40–61). Hershey, PA: IGI Global. doi:10.4018/978-1-5225-0654-6.ch003

Shi, J., Erdem, E., & Liu, H. (2016). Expanding Role of Telephone Systems in Healthcare: Developments and Opportunities. In A. Dwivedi (Ed.), *Reshaping Medical Practice and Care with Health Information Systems* (pp. 87–131). Hershey, PA: IGI Global. doi:10.4018/978-1-4666-9870-3.ch004

Shijina, V., & John, S. J. (2017). Multiple Relations and its Application in Medical Diagnosis. *International Journal of Fuzzy System Applications*, *6*(4), 47–62. doi:10.4018/IJFSA.2017100104

Shipley, N., & Chakraborty, J. (2018). Big Data and mHealth: Increasing the Usability of Healthcare Through the Customization of Pinterest – Literary Perspective. In J. Machado, A. Abelha, M. Santos, & F. Portela (Eds.), *Next-Generation Mobile and Pervasive Healthcare Solutions* (pp. 46–66). Hershey, PA: IGI Global. doi:10.4018/978-1-5225-2851-7.ch004

Singh, A., & Dutta, M. K. (2017). A Reversible Data Hiding Scheme for Efficient Management of Tele-Ophthalmological Data. *International Journal of E-Health and Medical Communications*, *8*(3), 38–54. doi:10.4018/IJEHMC.2017070103

Sivaji, A., Radjo, H. K., Amin, M., & Abu Hashim, M. A. (2016). Design of a Hospital Interactive Wayfinding System: Designing for Malaysian Users. In N. Mohamed, T. Mantoro, M. Ayu, & M. Mahmud (Eds.), *Critical Socio-Technical Issues Surrounding Mobile Computing* (pp. 88–123). Hershey, PA: IGI Global. doi:10.4018/978-1-4666-9438-5.ch005

Skourti, P. K., & Pavlakis, A. (2017). The Second Victim Phenomenon: The Way Out. In M. Riga (Ed.), *Impact of Medical Errors and Malpractice on Health Economics, Quality, and Patient Safety* (pp. 197–222). Hershey, PA: IGI Global. doi:10.4018/978-1-5225-2337-6.ch008

Smith, C. R. (2018). Medical Students' Quest Towards the Long White Coat: Impact on Mental Health and Well-Being. In C. Smith (Ed.), *Exploring the Pressures of Medical Education From a Mental Health and Wellness Perspective* (pp. 1–42). Hershey, PA: IGI Global. doi:10.4018/978-1-5225-2811-1.ch001

Smith, S. I., & Dandignac, M. (2018). Perfectionism: Addressing Lofty Expectations in Medical School. In C. Smith (Ed.), *Exploring the Pressures of Medical Education From a Mental Health and Wellness Perspective* (pp. 68–91). Hershey, PA: IGI Global. doi:10.4018/978-1-5225-2811-1.ch003

Sobrinho, Á. A., Dias da Silva, L., Perkusich, A., Cunha, P., Pinheiro, M. E., & Melo de Medeiros, L. (2016). Towards Medical Systems to Aid the Detection and Treatment of Chronic Diseases. In D. Fotiadis (Ed.), *Handbook of Research on Trends in the Diagnosis and Treatment of Chronic Conditions* (pp. 50–69). Hershey, PA: IGI Global. doi:10.4018/978-1-4666-8828-5.ch003

Soczywko, J., & Rutkowska, D. (2018). The Patient/Provider Relationship in Emergency Medicine: Organization, Communication, and Understanding. In A. Rosiek-Kryszewska & K. Leksowski (Eds.), *Healthcare Administration for Patient Safety and Engagement* (pp. 74–105). Hershey, PA: IGI Global. doi:10.4018/978-1-5225-3946-9.ch005

Sołtysik-Piorunkiewicz, A., Furmankiewicz, M., & Ziuziański, P. (2016). Web Healthcare Applications in Poland: Trends, Standards, Barriers and Possibilities of Implementation and Usage of E-Health Systems. In I. Deliyannis, P. Kostagiolas, & C. Banou (Eds.), *Experimental Multimedia Systems for Interactivity and Strategic Innovation* (pp. 258–283). Hershey, PA: IGI Global. doi:10.4018/978-1-4666-8659-5.ch013

Soni, P. (2018). Implications of HIPAA and Subsequent Regulations on Information Technology. In M. Gupta, R. Sharman, J. Walp, & P. Mulgund (Eds.), *Information Technology Risk Management and Compliance in Modern Organizations* (pp. 71–98). Hershey, PA: IGI Global. doi:10.4018/978-1-5225-2604-9.ch004

Spinelli, R., & Benevolo, C. (2016). From Healthcare Services to E-Health Applications: A Delivery System-Based Taxonomy. In A. Dwivedi (Ed.), *Reshaping Medical Practice and Care with Health Information Systems* (pp. 205–245). Hershey, PA: IGI Global. doi:10.4018/978-1-4666-9870-3.ch007

Srivastava, A., & Aggarwal, A. K. (2018). Medical Image Fusion in Spatial and Transform Domain: A Comparative Analysis. In M. Anwar, A. Khosla, & R. Kapoor (Eds.), *Handbook of Research on Advanced Concepts in Real-Time Image and Video Processing* (pp. 281–300). Hershey, PA: IGI Global. doi:10.4018/978-1-5225-2848-7.ch011

Srivastava, S. K., & Roy, S. N. (2018). Recommendation System: A Potential Tool for Achieving Pervasive Health Care. In J. Machado, A. Abelha, M. Santos, & F. Portela (Eds.), *Next-Generation Mobile and Pervasive Healthcare Solutions* (pp. 111–127). Hershey, PA: IGI Global. doi:10.4018/978-1-5225-2851-7.ch008

Stancu, A. (2016). Correlations and Patterns of Food and Health Consumer Expenditure. In A. Jean-Vasile (Ed.), *Food Science, Production, and Engineering in Contemporary Economies* (pp. 44–101). Hershey, PA: IGI Global. doi:10.4018/978-1-5225-0341-5.ch003

Stanimirovic, D. (2017). Digitalization of Death Certification Model: Transformation Issues and Implementation Concerns. In S. Saeed, Y. Bamarouf, T. Ramayah, & S. Iqbal (Eds.), *Design Solutions for User-Centric Information Systems* (pp. 22–43). Hershey, PA: IGI Global. doi:10.4018/978-1-5225-1944-7.ch002

Stavros, A., Vavoulidis, E., & Nasioutziki, M. (2016). The Use of Mobile Health Applications for Quality Control and Accreditational Purposes in a Cytopathology Laboratory. In A. Moumtzoglou (Ed.), *M-Health Innovations for Patient-Centered Care* (pp. 262–283). Hershey, PA: IGI Global. doi:10.4018/978-1-4666-9861-1.ch013

Sukkird, V., & Shirahada, K. (2018). E-Health Service Model for Asian Developing Countries: A Case of Emergency Medical Service for Elderly People in Thailand. In M. Khosrow-Pour (Ed.), *Optimizing Current Practices in E-Services and Mobile Applications* (pp. 214–232). Hershey, PA: IGI Global. doi:10.4018/978-1-5225-5026-6.ch011

Sygit, B., & Wąsik, D. (2016). Patients' Rights and Medical Personnel Duties in the Field of Hospital Care. In A. Rosiek & K. Leksowski (Eds.), *Organizational Culture and Ethics in Modern Medicine* (pp. 282–297). Hershey, PA: IGI Global. doi:10.4018/978-1-4666-9658-7.ch012

Sygit, B., & Wąsik, D. (2016). The Idea of Human Rights in Conditions of Hospital Treatment. In A. Rosiek & K. Leksowski (Eds.), *Organizational Culture and Ethics in Modern Medicine* (pp. 236–254). Hershey, PA: IGI Global. doi:10.4018/978-1-4666-9658-7.ch010

Talbot, T. B. (2017). Making Lifelike Medical Games in the Age of Virtual Reality: An Update on "Playing Games with Biology" from 2013. In B. Dubbels (Ed.), *Transforming Gaming and Computer Simulation Technologies across Industries* (pp. 103–119). Hershey, PA: IGI Global. doi:10.4018/978-1-5225-1817-4.ch006

Tamposis, I., Pouliakis, A., Fezoulidis, I., & Karakitsos, P. (2016). Mobile Platforms Supporting Health Professionals: Need, Technical Requirements, and Applications. In A. Moumtzoglou (Ed.), *M-Health Innovations for Patient-Centered Care* (pp. 91–114). Hershey, PA: IGI Global. doi:10.4018/978-1-4666-9861-1.ch005

Tiago, M. T., Tiago, F., Amaral, F. E., & Silva, S. (2016). Healthy 3.0: Healthcare Digital Dimensions. In A. Dwivedi (Ed.), *Reshaping Medical Practice and Care with Health Information Systems* (pp. 287–322). Hershey, PA: IGI Global. doi:10.4018/978-1-4666-9870-3.ch010

Tripathy, B. (2016). Application of Rough Set Based Models in Medical Diagnosis. In S. Dash & B. Subudhi (Eds.), *Handbook of Research on Computational Intelligence Applications in Bioinformatics* (pp. 144–168). Hershey, PA: IGI Global. doi:10.4018/978-1-5225-0427-6.ch008

Turcu, C. E., & Turcu, C. O. (2017). Social Internet of Things in Healthcare: From Things to Social Things in Internet of Things. In C. Reis & M. Maximiano (Eds.), *Internet of Things and Advanced Application in Healthcare* (pp. 266–295). Hershey, PA: IGI Global. doi:10.4018/978-1-5225-1820-4.ch010

Unwin, D. W., Sanzogni, L., & Sandhu, K. (2017). Developing and Measuring the Business Case for Health Information Technology. In K. Moahi, K. Bwalya, & P. Sebina (Eds.), *Health Information Systems and the Advancement of Medical Practice in Developing Countries* (pp. 262–290). Hershey, PA: IGI Global. doi:10.4018/978-1-5225-2262-1.ch015

Urooj, S., & Singh, S. P. (2016). Wavelet Transform-Based Soft Computational Techniques and Applications in Medical Imaging. In P. Saxena, D. Singh, & M. Pant (Eds.), *Problem Solving and Uncertainty Modeling through Optimization and Soft Computing Applications* (pp. 339–363). Hershey, PA: IGI Global. doi:10.4018/978-1-4666-9885-7.ch016

Vasant, P. (2018). A General Medical Diagnosis System Formed by Artificial Neural Networks and Swarm Intelligence Techniques. In U. Kose, G. Guraksin, & O. Deperlioglu (Eds.), *Nature-Inspired Intelligent Techniques for Solving Biomedical Engineering Problems* (pp. 130–145). Hershey, PA: IGI Global. doi:10.4018/978-1-5225-4769-3.ch006

Waegemann, C. P. (2016). mHealth: History, Analysis, and Implementation. In A. Moumtzoglou (Ed.), M-Health Innovations for Patient-Centered Care (pp. 1-19). Hershey, PA: IGI Global. doi:10.4018/978-1-4666-9861-1.ch001

Watfa, M. K., Majeed, H., & Salahuddin, T. (2016). Computer Based E-Healthcare Clinical Systems: A Comprehensive Survey. *International Journal of Privacy and Health Information Management*, *4*(1), 50–69. doi:10.4018/IJPHIM.2016010104

Wietholter, J. P., Coetzee, R., Nardella, B., Kincaid, S. E., & Slain, D. (2016). International Healthcare Experiences: Caring While Learning and Learning While Caring. In D. Velliaris & D. Coleman-George (Eds.), *Handbook of Research on Study Abroad Programs and Outbound Mobility* (pp. 470–496). Hershey, PA: IGI Global. doi:10.4018/978-1-5225-0169-5.ch019

Witzke, K., & Specht, O. (2017). M-Health Telemedicine and Telepresence in Oral and Maxillofacial Surgery: An Innovative Prehospital Healthcare Concept in Structurally Weak Areas. *International Journal of Reliable and Quality E-Healthcare*, *6*(4), 37–48. doi:10.4018/IJRQEH.2017100105

Wójcik, A., & Chojnacki, M. (2016). Behavior and Ethical Problems in the Functioning of the Operating Theater (Case Study). In A. Rosiek & K. Leksowski (Eds.), *Organizational Culture and Ethics in Modern Medicine* (pp. 154–179). Hershey, PA: IGI Global. doi:10.4018/978-1-4666-9658-7.ch007

Wolk, K., & Marasek, K. P. (2016). Translation of Medical Texts using Neural Networks. *International Journal of Reliable and Quality E-Healthcare*, *5*(4), 51–66. doi:10.4018/IJRQEH.2016100104

Wong, A. K., & Lo, M. F. (2018). Using Pervasive Computing for Sustainable Healthcare in an Aging Population. In B. Fong, A. Ng, & P. Yuen (Eds.), *Sustainable Health and Long-Term Care Solutions for an Aging Population* (pp. 187–202). Hershey, PA: IGI Global. doi:10.4018/978-1-5225-2633-9.ch010

Wu, W., Martin, B. C., & Ni, C. (2017). A Systematic Review of Competency-Based Education Effort in the Health Professions: Seeking Order Out of Chaos. In K. Rasmussen, P. Northrup, & R. Colson (Eds.), *Handbook of Research on Competency-Based Education in University Settings* (pp. 352–378). Hershey, PA: IGI Global. doi:10.4018/978-1-5225-0932-5.ch018

Yadav, N., Aliasgari, M., & Poellabauer, C. (2016). Mobile Healthcare in an Increasingly Connected Developing World. *International Journal of Privacy and Health Information Management*, *4*(2), 76–97. doi:10.4018/IJPHIM.2016070106

Yadav, S., Ekbal, A., Saha, S., Pathak, P. S., & Bhattacharyya, P. (2017). Patient Data De-Identification: A Conditional Random-Field-Based Supervised Approach. In S. Saha, A. Mandal, A. Narasimhamurthy, S. V, & S. Sangam (Eds.), Handbook of Research on Applied Cybernetics and Systems Science (pp. 234-253). Hershey, PA: IGI Global. doi:10.4018/978-1-5225-2498-4.ch011

Yu, B., Wijesekera, D., & Costa, P. C. (2017). Informed Consent in Healthcare: A Study Case of Genetic Services. In A. Moumtzoglou (Ed.), *Design, Development, and Integration of Reliable Electronic Healthcare Platforms* (pp. 211–242). Hershey, PA: IGI Global. doi:10.4018/978-1-5225-1724-5.ch013

Yu, B., Wijesekera, D., & Costa, P. C. (2017). Informed Consent in Electronic Medical Record Systems. In I. Management Association (Ed.), Healthcare Ethics and Training: Concepts, Methodologies, Tools, and Applications (pp. 1029-1049). Hershey, PA: IGI Global. doi:10.4018/978-1-5225-2237-9.ch049

Yu, M., Li, J., & Wang, W. (2017). Creative Life Experience among Students in Medical Education. In C. Zhou (Ed.), *Handbook of Research on Creative Problem-Solving Skill Development in Higher Education* (pp. 158–184). Hershey, PA: IGI Global. doi:10.4018/978-1-5225-0643-0.ch008

Zarour, K. (2017). Towards a Telehomecare in Algeria: Case of Diabetes Measurement and Remote Monitoring. *International Journal of E-Health and Medical Communications*, *8*(4), 61–80. doi:10.4018/IJEHMC.2017100104

Zavyalova, Y. V., Korzun, D. G., Meigal, A. Y., & Borodin, A. V. (2017). Towards the Development of Smart Spaces-Based Socio-Cyber-Medicine Systems. *International Journal of Embedded and Real-Time Communication Systems*, *8*(1), 45–63. doi:10.4018/IJERTCS.2017010104

Zeinali, A. A. (2018). Word Formation Study in Developing Naming Guidelines in the Translation of English Medical Terms Into Persian. In M. Khosrow-Pour, D.B.A. (Ed.), Encyclopedia of Information Science and Technology, Fourth Edition (pp. 5136-5147). Hershey, PA: IGI Global. doi:10.4018/978-1-5225-2255-3.ch446

Ziminski, T. B., Demurjian, S. A., Sanzi, E., & Agresta, T. (2016). Toward Integrating Healthcare Data and Systems: A Study of Architectural Alternatives. In T. Iyamu & A. Tatnall (Eds.), *Maximizing Healthcare Delivery and Management through Technology Integration* (pp. 270–304). Hershey, PA: IGI Global. doi:10.4018/978-1-4666-9446-0.ch016

Ziminski, T. B., Demurjian, S. A., Sanzi, E., Baihan, M., & Agresta, T. (2016). An Architectural Solution for Health Information Exchange. *International Journal of User-Driven Healthcare*, *6*(1), 65–103. doi:10.4018/IJUDH.2016010104

Zineldin, M., & Vasicheva, V. (2018). Reducing Medical Errors and Increasing Patient Safety: TRM and 5 Q's Approaches for Better Quality of Life. In *Technological Tools for Value-Based Sustainable Relationships in Health: Emerging Research and Opportunities* (pp. 87–115). Hershey, PA: IGI Global. doi:10.4018/978-1-5225-4091-5.ch005

# About the Contributors

**Harjit Pal Singh** is Associate Professor & Assistant Director (Research & Innovation) at CT Institute of Engineering, Management and Technology, Jalandhar. He has completed his Doctorate from National Institute of Technology, Jalandhar, Punjab. His area of interest is Intelligent Signal Processing, Voice over Internet Protocol, Wireless Sensor Networks, IOT, Artificial Intelligence, Wireless Communication, Computational Optimizations, Biomedical Signal Processing, Digital Filter Design, and Engineering Physics. His research and teaching experience is 12 years. His research findings have been published through more than 50 research papers published in referred/indexed journals and international conferences. He is the editor of 3 books with reputed international publishers such as Taylor & Francis, CRC Press, IGI USA etc. He has attended and presented his research findings in different countries such as France, UK, USA, Dubai, Hong Kong etc. He has chaired number of national/international conferences and also delivered expert talks on burning areas in signal processing, IoT, wireless sensor network etc. In academics and research, he successfully guided 2 PhD scholars and 5 PhD scholars are working under his supervision. At Master level, he has guided 29 Masters Dissertations and 5 Master students working under his supervision.

**Anurag Sharma** is doctorate from Dr B. R. Ambedkar National Institute of Technology, Jalandhar in the field of Artificial Intelligence for Health Care. He is M.Tech and B.Tech from Punjab Technical University, Jalandhar. He has more than 15 years of total experience in Teaching as well as in Industry. He worked with various industries like BEL, Punjab Communication Limited, Elcom Systems and in various institutions like DAVIET, Jalandhar, CT Group of Institutions, Jalandhar etc. He is presently working with GNA University as Professor in Faculty of Engineering, Design and Automation. He is recipient of ISTE best teacher's award thrice and is Wipro Mission 10x certified faculty member. He has guided more than 25 M.Tech students dissertation and is guiding 06 PhD scholars. He has published more than 80 research articles in reputed International journals/conferences. He has received grants from Indian Council of Medical Research (ICMR), Department of Science and Technology (DST) and other funding agencies for his research work.

**Vikas Khullar** is working as an Assistant Professor in the Department of Computer Science and Engineering at Chitkara University Institute of Engineering and Technology, Punjab, India. Dr Vikas received his doctorate degree in Computer Science and Engineering from IKG Punjab Technical University in 2020. The present area of his research is to develop assistive technologies to support individuals with neurological disorders. He is experienced in the area of intelligent, scalable and real time systems to develop assistive technologies. He has published 50 research papers in reputed journals and conferences.

* * *

**Mohammad Mudassir Ahmad** has completed his B.Tech degree in Electronics & Communications from DAV Institute of Engineering & Technology, Jalandhar, India affiliated to I K Gujral Punjab Technical University, Jalandhar, India. Currently, he is working as a Patent Research Analyst. He represents a review paper on Application of "Brain Computer And Brain Chip Interfacing" at the 22nd Punjab Science Congress, 2019. He is working on complex projects based on Arduino and Raspberry Pi. Currently, he is working on a review paper based on E-textile technology. His field of interest are Wireless Sensor Network, Optical Fibers, Automation, EMG based prosthetic arm, Lidar, Antenna, UWB & NFC, PCB Designing Layouts, Exoskeleton, Hydraulic system, IC Engines, Dark Energy, Dark Matter, Cosmic Web.

**Mohit Angurala** received his Ph.D. Degree in Computer Science and Engineering from IK Gujral Punjab Technical University. He completed his bachelor's degree in Information Technology from Beant College of Engineering and Technology in 2011. He received his master's degree in Computer Science Engineering and Technology from Punjab Technical University Main Campus in 2014. Now, he is working as Assistant Professor in computer science and engineering department in Chitkara University Rajpura Campus, Punjab. His research work includes energy enhancement in wireless sensor networks, mobile ad hoc networks, and network security.

**Strivathsav Ashwin Ramamoorthy** received the B Tech degree in Electrical and Electronics Engineering from the National Institute of Technology - Trichy in 2019. He has one year of professional experience of working in Oil and Gas industry and the healthcare sector. His research interests include medical image processing, computer vision and deep learning.

**Ashish Raman** is working as Assistant Professor, National Institute of Technology, Jalandhar. He has more than 15 years of teaching experience. His major area of interest includes VLSI and signal processing. He has published several papers in the International/National Journals/Conferences and book chapters.

# Index

www.ingramcontent.com/pod-product-compliance
Lightning Source LLC
LaVergne TN
LVHW081520300926
849161LV00006B/1234